Coping with Uncertainty

Kamil Maria Wielecki

Coping with Uncertainty

Petty Traders in Post-Soviet Russia

PETER LANG
EDITION

Bibliographic Information published by the Deutsche Nationalbibliothek
The Deutsche Nationalbibliothek lists this publication in the Deutsche Nationalbibliografie; detailed bibliographic data is available in the internet at http://dnb.d-nb.de.

Cover Illustration courtesy of Perach Ben Chaim.

The work was prepared within the project "The Traditions of Mediterranean Humanism and the Challenges of Our Times: the Frontiers of Humanity", supported by the Foundation for Polish Science — International PhD Programme, co-financed by the European Union within the European Regional Development Fund.

ISBN 978-3-631-66855-9 (Print)
E-ISBN 978-3-653-06135-2 (E-Book)
DOI 10.3726/978-3-653-06135-2

Peter Lang – Frankfurt am Main · Bern · Bruxelles · New York · Oxford · Warszawa · Wien

This publication has been peer reviewed.

www.peterlang.com

For Kinga

Contents

Acknowledgments 9

Note on transliteration 11

1. Introduction: surviving in the uncertain post-Soviet world 13

Ordering the disorder: new regulations on retail trade in Russia 16

The roots and fruits of petty trade: uncertain past, uncertain present 21

2. Understanding uncertainty and defining *neopredelyonnost'* 25

Uncertainty and entrepreneurship in Russia 29

The economic man vs. the socially embedded man 30

3. From informal Soviet economy to Russian petty trade 35

Researching the socialist informal economy 36

The everyday world of real socialism: Aleksandr Zinovyev and his *homo sovieticus* 41

The kolkhoz market and *barakholka*: petty trade in socialism 44

Mikhail Gorbachev and perestroika 47

A passage between epochs: the late Soviet entrepreneurship 52

4. Coping with uncertainty: surviving the Russian transformation 55

Incomprehensible to mortals: how could one manage in the 1990s? 58

The *chelnoki* 67

The women who saved Russia 75

Entrepreneurship, informality and criminality in Russia 77

5. **Embedding economy at Russian open-air markets** 89
Living in legal uncertainty: the Slavyansky market epic 91
Coping with uncertainty in workplace: the Slavyansky traders 93
Ethnic economy and trade minorities 102
Building communities at Russian open-air markets 113

6. **Conclusions: an insight into the political economy of Russia** 117

Appendix 1. Methodology 125

Appendix 2. List of Interview Partners 135

Appendix 3. Voices from open-air markets 141

Works cited 153

Photographies 169

Index 179

Acknowledgments

This book would have not been created if I had not received selfless support from many people. At a fundamental level, an ethnography is, as a matter of fact, a collective endeavor – for it is possible due to the willingness of interview partners to share their experience with an anthropologist. Since their anonymity must be protected for ethical reasons, I can only express general gratitude to all those who contributed to this work with their words and deeds.

Apart from interview partners, there are people – especially scholars – whose names should be mentioned. First of all, this pertains to Jan Kieniewicz, who has never spared his time to talk with me. Our conversations served for me as a great source of inspirations for posing questions and looking for answers. Special thanks should go to a number of Polish and Russian scholars – Jan Miernowski, Ivan Peshkov, Irina Savelieva, Vadim Radaev, Eugenia Bukharova, Viktor Dyatlov and Yuri Popkov – who during my studies at the University of Warsaw as well as my trips to Russia discussed with me issues of my work, broadening my view on petty trade and Russian systemic transformation. I would likewise like to thank Jakob Rigi and Violetta Zentai for their elucidating observations and many practical hints concerning doing anthropology in a postsocialist context. I am also grateful to Michał Buchowski and Szymon Wróbel for their careful reading of my work and invaluable remarks. I would like to thank Jakub Ozimek for editing this book and the countless critical, yet constructive comments he made. Cristopher Culver's proofreading was also very helpful in preparing this text for publishing.

I am grateful to my host institution – the Faculty of Artes Liberales (University of Warsaw) – for offering me a perfect intellectual space and to the Foundation for Polish Science for funding my work. I also would like to express my gratitude to the Higher School of Economics (Moscow) and the Central European University (Budapest) for granting me scholarships which substantially contributed to the completion of my project.

Finally, I would like to thank my wife Kinga for her constant, and absolutely indispensable, support.

K. M. W.

Warszawa, June 2015

Note on transliteration

I use the Romanization style for Russian set by the US Board on Geographic Names (BGN) and the Permanent Committee on Geographical Names (PCGN), known also as the British Standard. I adopted a simplified version of this style, rendering *ё* as *yo* and shortening the *-iy* and *-yy* endings to *–y*. For the soft sign *ь* I use the apostrophe ʻ and for the hard sign *ъ* the quotation mark ". As far as different styles of Romanization are in use, some names might occur in different versions, e.g. *Radayev* and *Zinovyev* might occur as *Radaev* and *Zinoviev* in bibliographical references. I make some exemptions from these rules for the sake of familiarity. For instance, instead of *Gorbachyov* and *Gaydar*, I write *Gorbachev* and *Gaidar.*

1. Introduction: surviving in the uncertain post-Soviet world

Nadya, a tradeswoman from Krasnoyarsk, told me: *As far as the state goes, well, it just abandoned us like a bunch of blind kittens. Nobody cared about us not receiving our salaries... Everything we produced in the Soviet Union – it all became useless and it all disappeared in an instant. And it was so difficult, you cannot imagine...* (T5).[1] She was describing the situation of Russia in the 1990s, where millions of people, having all been suddenly dismissed from state enterprises, were forced to survive on their own. What the newborn country faced after the dissolution of the Soviet Union was in fact a huge civilizational crisis. The policy of shock therapy and attempts at developing the market economy in the country resulted in an economic collapse. State property was massively expropriated, inflation raged, and Russian began to be haunted by unemployment. But the crisis went far beyond the economy. The depression of the state led also to the collapse of social services, a dramatic decline in life expectancy, extreme lawlessness, and a rise in corruption and violent crime.

Nevertheless, as Michael Burawoy put it, "somehow Russian people have managed to sustain a day to day existence despite the cataclysm depicted in official figures" (1999: 1). This study will show how this "somehow" was achieved in practice.

Nadya, for instance, became a *chelnok* (pl. *chelnoki*), i.e. a shuttle trader, and she travelled to Poland in order to buy goods and sell them at home. She also learned how to sew fur hats, which she then sold at an open-air market. Her case may be treated as typical, since it was petty trade that allowed vast numbers of people to survive. The grave shortages of consumer goods appeared already by the end of the Soviet era. Then, in the 1990s, they only deepened as Russian industry collapsed. The workplaces of many people simply vanished, but even those who still worked could go unpaid for months or years on end, and they had to earn their living some other way. Others, in their turn, received their salaries in products made by the enterprises they worked in. They were thus forced to exchange these products or sell them. All these factors pushed masses of Russian citizens to engage in trade.

This engagement took on various forms. Some people just sold what they possessed, such as books. Others focused on reselling: along the Trans-Siberian Railway people waited for Beijing-Moscow trains, which at that time came packed with Chinese clothes. They then sold the clothes directly in the streets. There were also people who opened their own kiosks in the country and bought wares from wholesalers in local urban centers. The most popular form, however, came to be international shuttle trading. As soon as the end of 1991, people spontaneously started to go abroad in order

1 The symbols in the brackets after a name or at the end of a quote refer to particular interview partners. A list of them, which contains short descriptions of all quoted persons, can be found in Appendix 2. In order to ensure the anonymity of the people I talked to, I changed their names.
All interviews were conducted in Russian. All quotations from the interviews were translated into English either by myself (K. M. W.) or by Jakub Ozimek (see Appendix 3). All translations of Russian terms and quotations, if not indicated others in the Works Cited section, were done by me (K. M. W.).

to obtain goods and sell them at home. Poland, Turkey, and China were their most popular destinations, and – since they came from literally every Russian region – in order to get abroad, they travelled sometimes thousands of kilometers, mainly by means of buses, trains and ships. In the first years of the systemic changes, there were neither professional cargo companies nor a system of deliveries, so everything that *chelnoki* acquired, they basically had to carry themselves. Then they traded their goods at open-air markets. In the mid-1990s, according to various estimates, there were from 18 up to even 30 million people engaged in international shuttle trade (cf. Kapralova and Karaseva 2005: 402; Mel'nichenko 1997: online). The numbers were thus colossal, as was the quantity of goods they imported into Russia.

Since the 1990s, petty trade in Russia has undergone a significant evolution. The *chelnoki* era may be regarded to have ended with the financial default of August 1998. It was then that the ruble lost its value in relation to the dollar by several times. Everyone who had any financial obligations in foreign currencies went bankrupt immediately. That included banks as well as individual traders who made their dealings with foreign partners almost entirely in dollars. Granted, there were some shuttle traders who managed to survive in spite of all the difficulties. There were also people who started their businesses after the crisis. All in all, however, the quantity of people involved in trade decreased substantially. In addition, the phenomenon remained international only in the borderlands whereas most of the traders were based. If traders go somewhere to get the goods, they would go to Russian urban centers. One no longer has to go abroad in order to get supplies. Nevertheless, hundreds of thousands of people throughout Russia still rely on petty commerce to provide for themselves and their families.

In general, petty traders can be counted among the masses of people in the whole post-Soviet region who became dispossessed as a consequence of the systemic transformation.[2] They were deprived of their property, jobs and ascribed places in society. The dispossessed constitute a broad category insofar as they comprise such different groups as homeless people, economic migrants, or abandoned pensioners. However, what unites all of them is that they lost the material and social status they enjoyed in the socialist past. The feeling of loss and of being poor (*bednye*) is to a considerable extent subjective and is strengthened by the comparison to those who managed to take advantage of the changes (Humphrey 2002b: 21ff.; Nazpary 2002: 13–16).

---

2 It was a matter of vehement discussions whether the changes in Central and Eastern Europe (CEE) should be called a *transition* or *transformation*. Personally, I find it much better to speak of postsocialist or systemic *transformations* than of *transitions* to capitalism or to a market economy in CEE. As has been argued, *transition* implies a colonialist teleological assumption of Western-style capitalism as the final arrival point of the changes. Such an approach raises a question: from what kind of socialism does one depart and to what kind of capitalism should one arrive? For just as there were various socialisms, there is no universal capitalism either (cf. Bryant and Mokrzycki 1994: 2–4; Hann and Hart 2011: 136). Furthermore, I would like to avoid the trap of what Michał Buchowski called *transitology*, a neoliberal and highly ideological discourse that created *new others* of the transitions. These *others* were groups that were not able to follow the rules of the radically changed social reality and therefore were blamed for being lazy and civilizationally incompetent (Buchowski 2001: 14–17).

The vendors – just as the whole category of the dispossessed – originated from different social classes. People of all walks of life were involved in trade at open-air markets in the 1990s, ranging from pensioners, members of ethnic minorities, or Russian refugees from the former Soviet republics to university professors. Overall, however, it was former industrial workers and the so-called working intelligentsia who engaged into trade in order to survive. Furthermore, as it has been argued, one has to use ideas concerning classes and social stratification in postsocialist countries cautiously (Buchowski 2001: 141–143; Il'in 2000: 86–87; Il'in 2008). Even if petty traders in contemporary Russia were to be seen as forming a distinct social class, it would encompass several subclasses, delineated along the lines of entrepreneurs versus sellers, local tradespeople versus labor immigrants, stallholders versus employees and self-employed persons, etc. However, in a neoliberal regime "the new precarious poverty grows in an atmosphere of silence" (Beck and Beck-Gernsheim 2002: 52). Unlike in previous periods, the fact of being poor does not carry any budding dignity. On the contrary, it is perceived as one's personal failure and a sign of helplessness. Thus, in modern societies, poverty has lost is class-distinguishing potential and instead rather supports social atomization.

A few external factors seem to be shaping the actual state of petty trade in Russia at the moment. The main challenge for the people engaged in the so-called street economy (Ivleva 2010) – i.e. those who work at open-air markets, in street kiosks, or trade informally in the streets – is the competition posed by big shopping centers and rapidly developing supermarket chains. The costs of economic crises are also suffered by the vendors: the prices of goods have gone up and at the same time the people have less money to spend. Last but not least, Russian cities are destinations of migrant workers (*trudovye migranty*) from the former Soviet republics (cf. Ivleva 2009). A large inflow of Chinese people can also be observed. Entire branches of trade and professions have become monopolized by members of a single nation, to such an extent that in some remaining markets it is the Russians[3] who constitute the minority. This inflow creates a great deal of tension between the newcomers and the local population in general, and poses a harsh competition for local traders in particular.

There is, however, one more factor which is absolutely decisive for the situation of petty trade in contemporary Russia. This factor is of a legislative nature, and as it is entangled in many other problems it requires a separate discussion.

---

3 I use the term *Russians* (*russkiye*) in a general meaning, in accordance with how it is popularly understood, that is, as a kind of label denoting not only ethnic Russians, as the term is lexicographically defined, but all of the Slavic or even white population of Russia. An additional feature of Siberia is that its inhabitants are usually of very mixed descent, combining Russian, Ukrainian, German, Polish, Jewish, and other ancestors so that – as they say – pure Russians do not exist there. Furthermore, one should note that there are basically no indigenous people in Krasnoyarsk anymore, so to a great extent the term *Russians* corresponds to the population of the city. Nevertheless, due to labor immigration and the general multinational character of the Russian Federation I prefer to use the term *Russian citizens* or *citizens of Russia* (*rossiyane*), for it describes all inhabitants of Russia regardless of their ethnicity.

Ordering the disorder: new regulations on retail trade in Russia

In 2006, the new act *On retail markets and on introducing changes to the employment code of the Russian Federation* (*O roznichnykh rynkakh i o vnesenii izmeneny v Trudovoy kodeks Rossiyskoy Federatsii*) was passed by the State Duma. The law was then repeatedly discussed and underwent a series of changes. Eventually, it was signed by President Dmitry Medvedev in December 2011 and came into force on January 1, 2013. The act imposed several new regulations on how a retail market should be equipped and organized in terms of space and functioning. The biggest change it introduced, however, was the prohibition on the use of so-called temporary structures (*vremennye sooruzheniya*) in retail markets (cf. *Zakon o roznichnykh rynkakh* 2006, art. 24, par. 3). This meant that virtually all open-air markets throughout the entire country would have to be closed or thoroughly renovated, since trade there was performed not in fixed buildings but in stalls, tents, containers, booths, etc. As a matter of fact, it brought an end to petty trade in the form it had taken during its boom in the 1990s.[4]

The 1990s were a time when open-air markets in Russia were created, usually in a spontaneous manner. They would appear at defunct sport facilities, in abandoned buildings, on empty squares or simply in any public place where streams of people were passing by. At a certain moment some traders would appear at such a place, selling their goods from camp beds or makeshift tents. They were likely to attract customers, since many kinds of products were scarce at that time and they offered attractive prices in comparison with regular shops. Once interest was established, other traders would join in. Within a few years, such a market could develop into a large structure.

In Russian, this way of organizing an open-air market is called *stikhiyny* "elemental". *Stikhiynoye bedstvie* is, for instance, a *natural disaster*. This epithet reflects the public attitude towards petty trade as something wild or uncontrolled. In post-Soviet society, trading was viewed as disorder-producing (Humphrey 2002b: 93–95). A negative stereotype of the private entrepreneur in general, and the petty vendor in particular, derived from the Soviet era, when private trading was declared an act of speculation and as such was heavily condemned by the state. In the 1990s, additional factors only strengthened this pejorative view. Common perceptions linked street vendors with smugglers and cheats. In addition, bazaars offended the public not only with their "elemental" architecture, as they consisted of rather chaotic collections of stands, booths and kiosks – but they were also seen as leading to crime. This was well rooted in reality, as open-air markets were indeed usually controlled by various mafia groups and even an old woman who wanted to sell tomatoes on the pavement usually had to pay a mafia protector for the privilege. Last but not least, the street economy attracted labor migrants from the former Soviet republics and China. Those migrants, especially those from the Caucasus (*kavkaztsy*), were both despised and feared.

In her famous definition, Mary Douglas defines dirt as something that offends against order (1996: 2). Dirt is something out of place, something that is at odds with established rules and values. So public opinion thought of workers who traded instead of producing, pensioners forced to sell their property, and economic migrants, or war

4 In the case of fruit and vegetable markets (*sel'skokhozyaystvennye rynki*) at which domestic producers sell their produce the law would come into force 2 years later, on January 1, 2015 (cf. *Zakon o roznichnykh rynkakh* 2006, art. 24, par. 4).

refugees who had left their countries in search of livelihood. Seedy open-air markets, which mushroomed in public spaces, were included in such pejorative perceptions as well. People trading in the streets became an icon of the new unruly times that came and replaced a well-organized past. The notions from popular language – words such as *bespredel* (*limitlessness, lawlessness*) and *bardak* (*brothel, disorder, disintegration*) that were used as *de facto* synonyms – expressed this feeling: the new times are characterized by injustice suffered by masses of people, their social degradation, collective corruption, and an shocking rise in violence. The two notions point to the chaos that followed in the aftermath of socialism (cf. Lindquist 2006: xiii–xiv; Nazpary 2002: 2).

Accordingly, since the early 1990s, different bureaucratic means were undertaken to establish order in the sphere of petty trade. Since the dissolution of the USSR, Russian open-air markets witnessed many raids performed by the police and other law enforcement structures in search of illegal immigrants and people trading without obligatory licenses. Street vendors have often been fined, and markets were renovated or closed. The act *On retail markets...* can be seen as a capstone of this tendency. State officials of lower or higher rank explained that the aim of the act is to make trade "well-ordered", "more cultured", or "more civilized" (cf. Minpromtorg 2013: online; Moshkov 2013: online; PenzaNews 2011: online). This meant above all the tightening of sanitation and security norms, but the act's influence reached far beyond that.

It was neither the necessity to obey the sanitation norms, which the act stipulated, that disturbed trades people the most, nor the general improvement of work conditions it required. One must remember that in the Siberian climate, trading at an open-air market means working in the freezing cold a few months a year. Only in recent years have vendors started to upgrade their stalls to kiosk buildings, but many of them still work outside the entire year. Tradespeople would therefore not mind being employed in a place that is heated and has access to running water. Nor would they resist making their workplaces safer and securer in terms of being more protected from fire or criminal activity. What disturbed tradespeople was the attitude of the authorities towards them and underprivileged classes in general: people who, after all, were trying to sustain their families in the honest way.

Turning to Michel Foucault's notion of *governmentality*, defined as the use of political economy as a rationale for exercising power by means of a security apparatus (cf. Foucault 1991b: 102–103), one might observe that in this instance of governmentality, the civilizational claim was just a pretext to expand more control over trade. From a more narrow perspective, the act *On retail markets...* aimed at further regulating labor immigration. This concerned especially migrants from the Caucasus and Central Asia, who massively took up jobs in petty trade, often violating Russian immigration laws (cf. Jarzyńska 2014: online; PenzaNews 2011: online; Protsenko 2013: online). In this respect, the act developed earlier legislative solutions that aimed at reducing the amount of foreigners working in retail trade. In the new millennium state authorities intended, on one hand, to encourage employing members of the local population and to support national entrepreneurs and, on the other, to persuade foreign entrepreneurs to try to obtain Russian citizenship. Since 2008 Russian citizenship has become necessary to find employment in trade or construction. In practice, however, it is enough to have a temporary residence permit (*propiska*) to work at a market.

From a broader perspective, the act *On retail markets...* aimed at changing the character of petty trade in general. It entailed, first of all, reducing the quantity of all tradespeople, regardless of their citizenship. Only some of them were able to move their businesses from temporary structures into shopping centers, as this required taking up new challenges, including securing much more financial means than when working at a market. What is more, the act has led to placing under closer surveillance those traders who have made the move into fixed buildings. A shopping center is much more transparent than an open-air market, and it is visibility which serves as one of the basic tools of governing (Foucault 1991a)

The law *On retail markets...* was successful in fulfilling its goals. Since 2007, when it was passed by the State Duma, outdoor markets have begun to be closed throughout Russia. As a consequence, the share of the markets in the overall turnover of retail trade decreased substantially. In Krasnoyarsk alone several markets were liquidated, especially in 2012–2013 (cf. *IA Regnum* 2013b: online; *Minpromtorg Rosii* 2013: online; Telegina 2012: online). As markets usually had attractive locations, housing estates or shopping centers were built in their place. With the closure of every market, from a few hundred up to a few thousand workplaces simply vanished. For instance, in the case of the Severny market in Krasnoyarsk around 400 workplaces were lost, but with the closure of the Gusinobrodsky in Novosibirsk – probably the biggest clothes market beyond the Urals – around 6,000 entrepreneurs alone lost their jobs (Komarov 2013: online; Telegina 2012: online; *IA Regnum* 2013a: online). The workplaces of producers, suppliers, delivery men, security guards and all people who provided markets with other services should be added to these numbers. According to various estimates, on average, one entrepreneur creates a workplace for one other person, which means that if an entrepreneur shuts down his activity, another workplace also vanishes with it (Avshalumova 2013: online).

Except for the tradespeople themselves, the discussed act had an impact also on their customers, all who were in search of cheaper products or better quality and had done their shopping from street vendors instead of going to malls. It also hit pensioners, to whom open-air markets offered free of charge so-called *social spots* (*social'nye mesta*), i.e. stalls to sell the produce from their gardens. Finally, it hit those local food producers for whom it did not make sense economically to rent a shopping space year-round, and were thus forced to use intermediaries.

Rents for the service sector – be it in a shopping center or any other fixed building – are obviously higher than those at an outdoor market. Therefore a trader must raise the prices of the goods he or she offers and it is eventually the customers who pay for the whole change. Thus, the initiative of Russian legislators to make petty trade well-ordered and more civilized actually caused collateral damage to a whole set of vulnerable social groups. That is why the closure of a market usually led to public protests. Petitions were signed and demonstrations organized by entrepreneurs and citizens who defended their local markets. Such protests did not usually bring any positive results, however (cf. *IA Regnum* 2013b: online). As one of the journalists commented with regard to the legislation, "When representatives of the state's legislative branch keep coming up with new plans to erect crystal palaces like the ones described by Chernyshevsky,[5] ordinary people always end up suffering badly. This is the tragedy of our society" (Protsenko 2013: online).

---

5 A reference to the famous social novel *What Is to be Done* by philosopher Nikolay Chernyshevsky. In the novel, Chernyshevsky presents a vision of a bright future for the human race. According to the vision, people will live in crystal palaces and

❖ Who can soften the law? Trade as a field of negotiation

A lot of markets throughout Russia have been closed. Nevertheless, some of them have somehow managed to survive or to prolong their existence far beyond January 1, 2013. In practice, the status of every single outdoor market was locally negotiated. This depended on many different factors, above all the position of its owners and their connections to local authorities. In this study, I describe in detail one case of such negotiation concerning the Slavyansky market in Krasnoyarsk. At the institutional level, the case offers what is, in my opinion, a fascinating insight into how politics is done in Russia, how government subjects and market actors are tied to each other. Nevertheless, besides looking at the interplay between different institutions that was evoked by the act *On retail markets...* one should also note the practical consequences that the law had for petty traders themselves.

One can call the situation of tradespeople who worked at open-air markets with unclear legal status as living in a legal uncertainty. This is the uncertainty of one's workplace and one's ability to cope with an unfavorable political economy. This proved to be the fate of vendors at many open-air markets throughout Russia. Nevertheless, such legal uncertainty affects not only markets. At every level of the street economy, from hawking clothes in an underpass to market administrators, people used some para-political practices to soften the legal requirements and to defend their positions. From such a perspective, the Russian battle for trade can be viewed as a field of relations of power, as the clashing and negotiation of different forces and values.

There was a little market by a metro station in the neighborhood where I lived in Moscow. As I was told by tradespeople working there, the market had in fact been closed several times, whether due to sanitation norms, immigration policies, or other reasons. The procedure was in fact the same each time: there were people to be bribed to issue new permits and there were higher rents to be paid. By the end of 2011 the monthly rents had reached 80,000 rubles (2,500 dollars) for a single booth, a circumstance that obviously translated into prices. These were higher than those offered in a hypermarket opposite the street, but the foodstuffs sold at the market were of better quality and still many people preferred to shop there.

Like at perhaps every market, there were traders who could not afford to rent a place. Again, their status was negotiated, too. There was a woman (T2) who used to sell from a booth but was then forced to abandon it, unable to pay the constantly rising rent. She stood with big bags and sold pickles and sauerkraut that she had produced. She and her female colleagues – who also traded informally in their own produce – were mercilessly persecuted by the police. As soon as they saw police officers approaching, they grabbed their goods and ran away. The women who sold clothes were force to do the same: they said they earned too little even to join together to bribe the officers. However, a man who informally sold honey that he produced (T3) was not, for whatever reason, afraid of the police, who allowed him to trade undisturbed. Although he himself did not want to elucidate this mystery to me, I was told by other tradespeople that he frequently presented the officers with a homemade light honey-based alcoholic beverage (*medovukha*) and that was the reason why they let him trade.

---

societies will be organized similarly to the Phalanstères proposed by Charles Fourier (cf. Chernyshevsky 1989 [1863]).

As one can see, other actors should also be named who have benefitted from imposing regulations on petty trade: public officials and representatives of various law enforcement bodies. Regulations concerning trade can be viewed as a part of the rise of the regulatory state, a characteristic tendency in Russia after the turn of the millennium (cf. Radaev 2011b: 5). There was often a specific interplay involved around the street vendors and the spontaneous markets. The latter – in keeping with their *elemental* nature – were like an ocean: they dispersed or were closed down and then they would reappear. On one hand, one can take the perspective of the traders – when survival of their families is at stake, people are ready to accept the risk of being fined and start to trade again. On the other hand, one should take into account the perspective of the officials, who are interested in being constantly needed, and to this end resort to creating new requirements. They do not want traders and individual entrepreneurs to perish completely. Rather, they are interested in sustaining a situation in which squeezing money out of these people is possible.

A question, which was heavily negotiated between officials and entrepreneurs in Krasnoyarsk, was the following: what does *a temporal structure* mean? The use of temporal structures in trade activity was forbidden by the act *On retail markets...* but the law did not provide a sufficiently clear definition of such a structure. Out of over 30 open-air markets which operated in the city in 2007, 6 years later only 15 were said to stand a chance of survival (cf. Telegina 2012: online; *Sibnovosti* 2014b: online). One way of fulfilling the legal requirements, and thus avoiding closure, was building a common roof over a market. Under such a renovation, all trading places were standardized in terms of size and appearance, but the actual conditions of trade did not usually change much. It was the roof that was important, because once it was built the market was treated as if it were inside a fixed building. Such renovation required, however, substantial financial resources from both the entrepreneurs and the market's administration and in practice it might prove insufficient for saving a market. For instance, it was claimed that the Tsentral'ny market might be closed even though it had a fixed building on its territory and was partially roofed. Its location in the city center made the market an attractive area for developers. Their claims to liquidate the market could be supported by legal arguments.

In summary, the changes the act has brought about are very telling. Overall, the legislative changes brought by the 2000s and 2010s determined the work activity as well as the personal lives of petty traders to a great extent. What was elemental in nature in the 1990s began, in the following decades, to be controlled and organized according to rules imposed by powerful actors in the Russian market. Legislation on retail trade can serve as a kind of a candle for casting some light on several further issues. Everything that has been happening around open-air markets, petty trade, and individual entrepreneurship in general for the last 20 years offers us an exceptional insight into several social problems of contemporary Russia, such as social disparities between the poor and the rich, unemployment, tensions between the center and the regions, immigration inflow and ethnic relationships in this multiethnic country. I will discuss these problems throughout this study.

The roots and fruits of petty trade: uncertain past, uncertain present

Taking into account the amazing progress petty trade had undergone in post-Soviet Russia – its main actors and results – I posed several research questions: how did it happen that in a country where private trade and entrepreneurship were basically forbidden for over 60 years such vast numbers of people became entrepreneurs virtually overnight? What were the survival strategies of the petty trader in the conditions of 1990s Russia? How do people cope with rapid social change? And more precisely: how do different individuals handle issues that concern everybody, such as hyperinflation, unemployment, poverty, and shortage of consumer goods? Apart from the questions that concerned predominantly the past, I wanted to tackle a few questions associated with the present: who are the petty traders now? What is their situation in contemporary Russia? What kind of relationships do they build with customers and among themselves? What are their main concerns, fears and hopes?

The studies on Russian transformation, or the many guises of postsocialist[6] transformations in general, have concentrated mainly on its political and macroeconomic aspects. Much less attention has been paid to the everyday life of so-called ordinary people. Considering the scale of the change and the size of the territory it afflicted, there is still a need for qualitative research concerning its impact on people's lives. It is here where this study can be placed: its aim is to look at the Russian systemic transformation from the perspective of petty traders. This entails an investigation into the ways the underprivileged classes of the society lived through the changes. I propose an anthropological or a grassroots perspective, which engages individual biographies as sources for the study of large social, economic, and political processes. This approach sees people as involved in history, which they co-create in their everyday practices. After all, it is the everyday actions of actual people which contribute to the formation of the reality we observe. It is here, not in abstract models or perceptions reconstructed in intellectual discourses, that we should look for inspiration for a fresh look at the transformation in Russia.

The empirical basis for this study is constituted by extensive ethnographic fieldwork that I carried out among tradespeople in the Siberian city of Krasnoyarsk in 2012–2013. My primary goal had been merely to provide an adequate description of the stories of traders, the problems they faced, the conditions in which they worked, and future perspectives. This seemingly humble goal turned out to be an ambitious one, because it required not only the reconstruction of petty traders' perspective

6 The term *postsocialism* may be perceived as a sign of a kind of helplessness in naming the reality in CEE and more generally all countries that have experienced a transition from socialism. As one of the authors put it over a decade ago, "Ten years after the collapse of the Soviet empire, one thing is clear: the word *postcommunism* has lost its relevance" (Rupnik 2002: 103). Nevertheless, the usage of the term seems reasonable to me, if only because of the temporary aspect: the socialist legacy is still meaningful in the former socialist countries. Moreover, it opens a wide field of comparison of how the transition from real socialisms to actually existing capitalisms took place (cf. Humphrey 2002a). In other words, the term is useful in comparing the past with the present. For discussions on theoretical perspectives of postsocialist studies, see: Brandtstädter 2007; Chari and Verdery 2009; Gille 2010; Hann 2002b.

of the transformation but also an analysis of the role that underprivileged classes played in the processes of change. Thus, achieving the goal would mean describing the whole of reality. And so, as every description of reality, this one was doomed to be incomplete, too. What I could do in order to elaborate a more complete picture of postsocialist transformations was to supplement the anthropological approach with general analyses provided by historians, economists, and sociologists. Therefore, the empirical material will be constantly collated with and checked against the existing scholarly literature on Russian transformation. In other words, I shall investigate the interrelations between micro and macro social processes and therefore combine both respective standpoints – the micro and the macro, the grassroots and the bird's-eye view, the individual and the national.

Contrary to what the ardent believers in shock therapy maintain, I argue – and this is the first main claim I make – that transformation can precede, influence, or even evoke transition. Following Jan Kieniewicz (2008: 175–191), I understand by *transition* a process of changing the political and economic institutions, while *transformation* points out people's ways of behavior oriented on contribution, resistance, or adaptation to changes. In this sense, *transformation* is broader and more important than *transition*, because it is actual people who create institutions and change the character thereof.

Behind the idea of subjecting Russia to shock therapy one can find a premise that one has to create capitalist institutions in order to break an alleged mentality of *homo sovieticus* and to change people's behavior. This meant social groups and individuals were considered to be more or less passive subjects who were supposed to just adapt to changes according to predominant patterns (cf. Buchowski 2001: 14–17; Buchowski 2006: 466–472; Kovács and Zentai 2012: 1–4). I, on the contrary, maintain that people started to change socialist institutions long before the transition was even launched. For example, the change happened partly due to traders and other people connected to the informal economy during socialist times. They invented quasi-market patterns of activity in entirely non-capitalist surroundings.

This can also be translated to the social, political, and economic processes in the 1990s. Indeed, various global processes – perestroika, the dissolution of the Soviet Union, the 1998 crisis and so on – were not started by petty traders. Such upheavals were imposed on them from above. Nevertheless, I argue that the traders possessed agency in shaping the course of the Russian transition and transformation. It was they who generated their own patterns of behavior and were capable of defending their own values. It was also their energy and resourcefulness that let the country get through the worst years of the crisis.

The second and more general claim I pose in this book is that during the 1990s as well as today, the lives of tradespeople and underprivileged classes in Russia in general have been characterized by uncertainty. This uncertainty encompasses many different dimensions including, among others, frequent turmoil haunting the Russian market, working conditions, and governmental policies towards individual entrepreneurship. In the 1990s you could not be certain about your savings, because inflation could make savings worthless or the national bank could declare its insolvency. You could not be sure if your pension money would be paid, because the state could withhold it. You could not trust that you had job security, because workplaces could be closed down on the grounds that they were unprofitable. You could not feel safe about your business, because you constantly faced the risk of extortion and harassment by criminals. Neither could you know whether the state you lived in would not dissolve and whether a war would not break out where previously peace prevailed. All these dimensions made up an existential uncertainty for tradespeople, whether Russian citizens or economic migrants, and for underprivileged classes in Russia in general. It was an uncertainty of what existence they are their families would have tomorrow. The whole Soviet world had collapsed and in coping with this situation people had to find other foundations of knowledge, certainty, and trust. While some have been successful, the lives of many are still wrought with instability.

❖ The content of the chapters

In the following chapter, entitled *Concepts and approaches*, I elaborate on the main concept of this study, namely that of uncertainty. Beginning with, and drawing on, the literature of the subject, I develop my own theory that aims at grasping the existential situation of Russian underprivileged classes in general and of tradespeople in Russia in particular. In this context, I also lay down in detail the interconnections between uncertainty, risk, and entrepreneurship, and I accentuate the active stance individuals can take while facing uncertainty, that is, the fact that they can cope with it in many possible ways. Next, I position myself within disciplinary fields. As the main tradition I follow is that of economic anthropology, by the very nature of this domain I will avail myself of developments from a few other disciplines.

In Chapter 3, *From the informal Soviet economy to Russian petty trade*, I tackle a paradox: how could a centrally-planned system which was said to support the citizens' passivity and collectivity become a breeding ground for individual entrepreneurship? I track down the preconditions of the outburst of petty trade in Russia in the 1990s in various informal economic practices developed during the Soviet times. I show that in the everyday world of real socialism an individual had to have skills usually attributed to entrepreneurs, such as risk-taking, innovation, and diligence. Otherwise, one could not solve basic problems in one's life. I also provide a historic and economic analysis of petty trade and private labor activities during the time of perestroika. What is important here is the connection of private entrepreneurship with informal and illegal phenomena. This heritage has burdened petty trade to this day and has constituted a source of negative social stereotypes.

In Chapter 4, *Coping with uncertainty: surviving the 1990s*, I focus my attention on the decade that is crucial for understanding not only the genesis of petty trade but also the whole social and political situation of Russia today. On the basis of my fieldwork material and other sources, I reconstruct the conditions that in the 1990s led to the outburst of petty trade. The central question of the chapter is: how do individual people cope with rapid social change? I start with an economic outline of reforms introduced by the first non-Soviet government of Yegor Gaidar. Then I go on to discuss the ways Russian people in the 1990s coped with their everyday problems: unemployment, the instability of the ruble, and the necessity of maintaining the household. A significant part of the chapter is devoted to the phenomenon of the *chelnoki*. I also provide a general analysis of the evolution petty trade in Russia underwent from the early 1990s up to the present day.

What fascinates me in this regard is people's resourcefulness, their particular variety of entrepreneurship. These skills allowed Russian citizens in general, and petty traders in particular, to withstand extremely difficult situations in the 1990s.

Their entrepreneurship has very much to do with survival. My interview partners frequently used different expressions attached to *life* (*zhizn'*) and *survival* (*vyzhivaniye*). *Life goes on!* was the joyful expression, with which Lyuba (T59) from the Slavyansky market would greet me. It meant that the market was still open and there were reasons to suppose it would continue to stay open at least for some time. It was this market, Lyuba kept repeating, that in the 1990s let people survive. For many, especially labor immigrants, survival was still a day-to-day struggle. The one-word sentence *vyzhivayem* ("we are surviving") could serve as an entire explanation of one's situation in life. In this chapter, I try to grasp the existential charge of such expressions.

In Chapter 5, *Embedding economy at Russian open-air markets*, I deal with the factors that shape the economic behavior of tradespeople at outdoor markets. I make a claim that traders' economic behavior is embedded in noneconomic factors, like ethnicity, religious beliefs, ties of friendship, etc. In the spirit of economic anthropology, I show how vendors challenge the economic models of the free market and perfect competition. I describe their actual behavior as being far from rational in purely pragmatic economic terms. All these factors cause competition to manifest itself in much softer forms than the paradigm of the individualistic *homo economicus* would assume. As a matter of fact, people often do cooperate and their economic strategies are shaped by more than merely a pursuit of profit.

I also show that traders at open-air markets form various kinds of communities. The easiest to build is a small and closed community based on ethnic ties. In such case a so-called trade minority appears. Such a community is characteristic especially of labor migrants. However, when people work together for a longer period of time and face common challenges, transnational communities also appear. My point is that many traders invest in social capital rather than just in economic capital. Traders cannot work solely on their own – they need other people whom they can trust. Traders also need to build trustful relationships with customers, for if customers feel just once they have been cheated, they will not come again. It is participation in various communities that helps vendors in coping with the uncertainty attached to their work activities.

2. Understanding uncertainty and defining *neopredelyonnost'*

A few years ago, the Slavyansky market in Krasnoyarsk was a busy place, where crowds of customers allowed one to make good money. In February 2012, when I visited the market for the first time, the situation looked quite different. The alleys at the market were to a large extent empty, as were many stalls. The former owners and lessees had abandoned them, being unable to run their businesses any longer. As those who remained explained to me, two supermarkets had opened in the vicinity, posing a serious competition to the market. An additional reason was the ensuing financial crisis. As traders had noticed, in recent years prices of foodstuffs had increased, while salaries inversely fell. As a result, customers simply had less money to spend. *This market is dying* was one of the first things I heard after I arrived.

However, neither competition nor crisis constituted the main source of the traders' anxiety. Because of legislative changes that limited outdoor trade, the Slavyansky was said to be up for closing. As matter of fact, however, the first rumors about the closure of the market had appeared a few years before, already in 2007. Different stories circulated among the tradespeople but nobody knew anything for sure. Since 2010, permissions to trade were issued first for 6, than for 3 months and eventually just from month to month. *The worst thing is the uncertainty* (*neopredelyonnost'*) I heard from Lyuba (T59), a tradeswoman whose story will be told in Chapter 4. She had an opportunity to supply herself with a batch of goods at an attractive price but she did not take advantage. She was afraid she would not sell them during the single month her permit covered. Uncertainty determined different decisions made by petty traders, but above all it concerned the most crucial of all issues: the future of the market.

The word used by Lyuba – *neopredelyonnost'* – means that something is undefined. In general, if you are uncertain, you do not know what is going to happen, you do not know how a situation is going to develop or how somebody crucial to your plans is going to act. As a result, you are uncertain about what you should do yourself. Uncertainty then is also about decision-making. What is undefined is the future, the consequences of your decisions and events that are independent of you, on which you do not have any influence. Sometimes you just wait in order to react properly to the course of events: you make predictions and have some suppositions, and so you can plan some strategies. In the worst case, however, you will be surprised because something that you had treated as certain turned out not to be so.

One can thus say that uncertainty is a state of limited knowledge, which is linked to decision-making and to the future. If this is so, then uncertainty has been universal for humankind throughout history. Obviously, there are times when we are 100% sure and simply know what to do. Many activities over the course of the day we perform mechanically, out of habit, and we do not have to consider them in any particular way. Nonetheless, there are still situations left in which a choice has to be made. As a matter of fact, we find ourselves in such situations quite often. Facing a choice, coping with uncertainty, is thus something common. This is the general premise subscribed to by some researchers of the subject – "uncertainty is everywhere and you cannot escape from it" (Lindley 2006: xi), "life is mostly made up of uncertainties" (Knight 1964 [1921]: 235).

Being a universal feature of human beings, modes of uncertainty nevertheless depend on the current social context. Uncertainty is subjective; it is always someone's uncertainty, which derives from individual perceptions and describes the relationship between a person and the world around them (Lindley 2006: 1). At the same time, these perceptions are more or less shared with other people from one's group, class, society, or culture (cf. Douglas and Wildavsky 1982: 193). And so, uncertainties are intersubjective and have also a collective dimension. Uncertainties are historically and socially embedded.

What, then, is so specific about *neopredelyonnost'*? If uncertainties vary depending on the historical and social context, it follows that instead of researching one universal uncertainty, one should investigate different modes in which it can appear. So what kind of uncertainty characterizes the Russian petty trader – or the Russian underprivileged classes in general – in the decades that followed the dissolution of the USSR?

One has to look for the defining features of *neopredelyonnost'* exactly in the form the Russian transformation took on and the point at which it started. One should keep the relational character of uncertainty in mind, and for Russian citizens it is basically the certainty of the Soviet past that serves as a point of reference. As Karl Polanyi argued, social calamities are primarily not of an economic but of a cultural nature – the industrial revolution "transformed within less than half a century vast masses of the inhabitants of the English countryside from settled folk into shiftless migrants (...) Not economic exploitation, as often assumed, but the disintegration of the cultural environment of the victim is then the cause of the degradation" (2001 [1944]: 164). In the same way, the collapse of socialism destroyed the cultural environment of masses of people, exposing them to extreme precariousness. As a matter of fact, it turned them too into shiftless migrants. This was the fate of the *chelnoki*, the Russians who escaped from the post-Soviet republics back to Russia out of fear of the new nationalisms, millions of other economic migrants from these republics, and generally all those forced to leave their homes because of the changes.

There was a process of vehement social individualization in the decades that have passed since the dissolution of the USSR. As Ulrich Beck and Ursula Beck-Gernsheim see it, individualization has to do with abolishing the ways of living established in the industrial society. With the passage to a risk society, such forms as recognized social stratification, class consciousness, and gender and family roles started to dissolve. Along with them, traditional modes of coping with anxiety and insecurity, which were embedded in one's social milieu, began to fail. Detraditionalized individuals are set free from the forms of the industrial society – one could also say they are dispossessed of these forms – and are left to cope with their problems by and large on their own. This liberation, however, turns out to be illusory, for the freedom it brings is quite precarious. It makes them dependent on the labor market and all domains connected with achieving success in it: education, consumption, social laws, fashion and so on (Beck 1992; 1999; Beck and Beck-Gernsheim 2002). In other words, citizens of the risk society fall into a new form of governmentality, that is, they are subjected to new ways of holding power over the population and the economy.

Independently of whether the Russian society has entered the stage of a risk society, the phenomena described above are visible if one looks at petty traders. People who had been employed in state enterprises and institutions became individual entrepreneurs, i.e. they started to work on their own. This also entailed work flexibilization – or for many coping with unemployment – and de-standardization of people's life courses. At the same time, however, the specificity of the Russian departure from the industrial society – or Russia's involutionary road to capitalism (Burawoy 2001) – stemmed from the initial state of the process, and more particularly from the type of entity an individual gets individualized from.

In Western capitalism, there was no entity comparable to the Soviet work collective. The collective constituted a basic social unit and its roles went far beyond one's professional life. It was through the collective, and one's workplace in general, that an individual became a fully-fledged citizen. Also, the distribution of goods, especially shortage goods, was carried out through the workplace. To a considerable extent, the collective also defined one's social position and functions. What is more, it had a significant impact on one's private sphere too, insofar as it organized people's communal life. This was due to many factors but the most important was the fact that members of one's collective were often his or her close friends and acquaintances. Thus, on the one hand, individuals depended strongly on their collectives and the collectives were powerful tools of extending social control. On the other hand, however, they constituted for an individual a safe and predictable environment. By belonging to a work collective, one secured one's existence.

Aleksandr Zinovyev, a Russian writer and philosopher, called the destruction of the work collective in the post-Soviet era a national tragedy. According to him, this destruction was not primarily due to the collapse of the Russian economy and the unemployment it brought. It was caused first of all by the privatization of state enterprises. Whereas a socialist workplace was oriented towards forming a work collective rather than gaining profit, in the new regime it was profit which became the key quality in assessing the success of an enterprise. All communal functions were given up as unprofitable. For Soviet citizens, their collectives were a kind of natural environment, the structures and people they could count on in case of need. Without her collective an individual was left to her own devices. "What they did to us is somewhat like taking a fish out of the water and into the shore and saying: There, now you are liberated from the communist water, now rejoice in the democratic shore! So here we are, rejoicing!" Thus says the narrator of Zinovyev's novel *Russian Tragedy* (2005: 25). Being withdrawn from their collectives, many Russians felt helpless and adrift. In this context petty traders may be called victims of decollectivization (cf. Hann and Hart: 2011: 132).

❖ Calculable risk and incalculable uncertainty

To simplify the issue, it can be stated that risk is a threat which does not exist until it happens. Nonetheless, the non-existent threat can affect our deeds and does so successfully when, for example, we take out an insurance policy. The institution of insurance assumes risk can be predicted and somehow calculated. Insurance not only converts possible events into capital, it does not stop at sheer possibility and pretends to secure the values of the present. In this sense, risk assessment constitutes a category of understanding, a part of a wider perception that contributes to the disenchantment of the world. It is a way of ordering reality which is founded on a presupposition that the uncertain future can be rendered in a calculable form and thus can be governed by human action. Such perception of risk – including making risk an object of scholarly analysis – is intrinsically a modern phenomenon (cf. Dean 1999: 131; Ewald 1991; Lindley 2006: 2; Weber 1948; Zinn 2008: 9–10).

A long scholarly tradition associates risk with entrepreneurship (cf. Cantillon 2010 [1755]: 73–77; Knight 1964 [1921]; Tarascio 1985: 251–252). Moreover, one can see a connection between the social perception of risk and the rise of capitalism. The first appearances of the notion of risk can be traced back to the transitional period between the Middle Ages and the early Modern era. It was then, for instance, that maritime insurance began to be used. The modern perception of risk lay in a realization that some results can be achieved only if something is at stake. One risks losing something in order to achieve something more – it is here that the field of entrepreneurial activity opens. The higher the stake the higher the profit should be, according to *high risk / high gain* principle. This is the basis on which the famous wager of Blaise Pascal rests: if what is at stake is salvation, unbelief does not make sense, for in comparison to the risk of damnation, it brings a disproportionately small profit (cf. Luhmann 1993: 9–11).

To what extent, however, can one predict the sinking of a ship? A merchant who sends his goods or orders transport from over the ocean can insure himself by buying a policy. Obviously, by doing so, he reduces his profit. What he gains in return is the certainty of not suffering an absolute loss. It is now up to the insurance agent to award compensation if the loss happens. What the agent does before issuing a policy is to assess the value of the transported goods, to check the route of the ship, to find out whether any pirates prowl these waters, and so on. In short, he uses statistical data on calamities and makes a probabilistic calculation of the possible foundering of this particular ship. The price of the insurance depends on this calculation.

Basically, two sorts of uncertainty can be distinguished in this example, namely risk and uncertainty proper. The major difference between these two categories is that risk is something that one can predict and measure. This does not apply to uncertainty, which is not a quantitative but rather an existential category. The two terms can also be called objective and subjective uncertainty, respectively. In the example of maritime trade, risk concerns the merchant and the statistical data. It is the merchant who has to assess the likelihood of the ship sinking and to include it into his calculation of profits and losses. This risk is likewise valid for an insurance agent, but in his case an additional factor comes into play: the uncertainty as to what will really happen. Will the ship call safely at port, will it sink, or will it perhaps be only damaged during a storm? The same uncertainty affects also the sailors on the ship.

For Frank Knight, who was perhaps the first to distinguish risk from uncertainty, only the latter defines a way of earning peculiar to the entrepreneur. Moreover, a substantial profit from entrepreneurial activity can be made only by taking up what he calls a true or proper uncertainty (Knight 1964 [1921]: 43–44, 231–232). The merchant who buys an insurance policy partially abdicates from his entrepreneurship; he turns into a banker who merely earns on interest rates. Interest earnings are certain – once you make a fixed-term deposit account, you are guaranteed to get your profit – but as a rule they have the disadvantage of being low. Risk and uncertainty are transferred to the insurance agent and thus it is the agent who becomes the true entrepreneur. It is the agent who, by risking a loss, will receive the undetermined residue or profit.

Indeed, in practice, calculable risk and incalculable uncertainty might be difficult to distinguish. Some authors have criticized the distinction as being too fuzzy and as obscuring all that lies in the middle: the practices of making the incalculable calculable and the other way around (Dean 1999: 132; Giddens and Pierson 1998: 103–105). As a matter of fact, there is a fair deal of notional confusion around the two terms. Many authors treat them as synonyms. For economists, engineers, medical doctors – in short, all those who can be placed within the techno-scientific approach – the problem is largely inexistent. What they investigate is risk treated as an objective and calculable phenomenon. However, they sometimes call it uncertainty, which only introduces more confusion. In most social science approaches, in turn, the difference is stressed, although usually with some reservations. Personally, I find the distinction useful and maintain it throughout this study.

What should be clarified here is that uncertainty does not preclude the possibility of predicting the future. Handling uncertainty may involve methods other than statistics and probability calculation, such as professional judgment, ordinary foresight, and rules of thumb. Incidentally, it is these uncertainty techniques that are usually utilized by entrepreneurs when they engage in the market. They base their decisions on their past experience, use practical judgment, accumulate information, and so on. Of course, one can call these techniques calculation of sorts, but what distinguishes them is that – unlike taking out insurance policies – they have the intrinsic feature of helping the people who engage in them in coping with uncertainty, not only with risks (O'Malley 2008: 72–73). Uncertainty concerns cases that are unique, and thus are not calculable, and in which individual, subjective sense-making plays a role.

Last but not least, uncertainty has one more crucial quality. Whereas risk describes a concrete danger, uncertainty concerns things that are essentially indefinite. In other words, the object of risk is much more clear. In that sense we speak about the risk of loss but the uncertainty of profit. One can conclude from that statement that uncertainty – as opposed to risk – need not pertain to a negative result but expresses general anxiety.

Uncertainty and entrepreneurship in Russia

As was mentioned above, uncertainty is a characteristic trait of an entrepreneur. Entrepreneurs – and particularly merchants or a traders, who can be perceived as prototypical figures of entrepreneurship – deals in their activity with risk and uncertainty. Risk constitutes a calculable projection of the future, i.e. what might happen and how it influences the budget of one's enterprise. Uncertainty, in turn, describes the margin that is left after all calculations are made, it describes the unpredictable. Only coping with this uncertainty – *true* uncertainty in Knight's terms – makes an entrepreneur a *true* entrepreneur. This does not mean he or she should not make any use of insurance policies. That could be seen as sheer gambling and as such would be unreasonable. It means that he or she has to find a middle way between taking perfectly safe steps and tackling the unknown. Only the latter might bring profits which are not a simple derivative of invested capital.

Nevertheless, my understanding of entrepreneurship is much broader than just running one's own business activity and coping with its peculiar uncertainties. Indeed, the majority of my interview partners – and the majority of tradespeople in general, I believe – are registered as individual entrepreneurs, i.e. self-employed persons. To limit my analysis to officially registered entrepreneurs, however, would be restricting. Firstly, during the 1990s, tradespeople did not have to register. Secondly, what I mean by entrepreneurship here lies not in having a status of a legal entity or in paying taxes in a specific regime. In the wide sense it is the ability to pull oneself up by one's bootstraps. It is a resourcefulness which entails taking responsibility for the material sustenance of one's family and finding alternative sources of income when no job seems to be available. In this sense, in 1990s Russia masses of dispossessed people demonstrated an entrepreneurial way of behavior. They decided what, where, and how much to do, although in many cases the possibilities of such choices were very limited.

You have to have some stable foundations for your knowledge. If something you have thought certain turns out not to be so, you can be taken aback. If the basic rules that govern everyday life cease to work, you become totally adrift. As a consequence of shock therapy, Russian citizens had no assurance of basic life continuity and frequently lost their foundations of certainty.

The new threats they had to face were of a very physical character and concerned the day-by-day subsistence of their families. The safety of work, of support from a collective, of currency, of the stability of the state – all the things that used to be certain in the Soviet past changed in the Russian present. There is a degree of uncertainty usually connected with trade activity: traders know their expenses but have only a rough idea of their gains. Traders do not know whether a customer will come and buy from them. A good that traders have invested in might rot and thus bring losses. A rival might make a more attractive offer and attract the customers that a trader has counted on, and so on. Apart from dangers of this kind, Russian petty traders had to cope with many other uncertainties concerning personal security, the overall economic situation, the implementation of legal regulations, and so on. Their uncertainty arose also from the rapid pace of changes. Reality was in constant flux, which precluded any distant planning (cf. Burawoy and Verdery 1999: 2). Although the changes are much slower today than they were in the 1990s, the basic uncertainty of life – as the cases of the Slavyansky and other open-air markets show – has remained similar. Now, however, it also has a new dimension. While in the 1990s the uncertainty of the future pushed masses of Russian people to engage into trade, it is the same uncertainty which draws millions of migrants to Russia today.

All these factors show that the uncertainty of life in post-Soviet Russia differs greatly from the uncertainties faced by individuals in Western welfare states, due most of all to the depth of Russian decline. Nevertheless, due to their resourcefulness, i.e. their ability to cope with uncertainty, tradespeople in Russia managed to survive even the worst periods of the crisis.

The economic man vs. the socially embedded man

The perspective which this book adopts in exploring petty trade in post-Soviet Russia is that of economic anthropology. At the outset of professional anthropology, researchers aimed at comprehensive descriptions of so-called primitive societies. In such endeavors, the economy could not be separated from other spheres of human life. One can see this clearly in Bronisław Malinowski's classical study on the institution of the Kula ring of Papua New Guinea: a system of exchanging necklaces and bracelets had to be considered in connection to social roles and stratification, political authority, and religious beliefs (2005 [1922]: 62–79). Later on, anthropologists turned to the investigation of modern societies. The field broadened but the scope of interest within it remained the same. In the case of economic anthropology, it is studying the interconnections between economy and other human activities. Economic behavior reaches in this perspective far beyond managing financial resources, it has to do with "how humans use the material world to maintain and express themselves in social groups" (Gudeman 2010: 210).

Exploring petty trade in post-Soviet Russia means in fact looking at the biographies, choices, and actions of people who engaged into trading activity. It is here where the difference between the approach adapted by classical economics and the approach of economic anthropology lies. The social sciences – and anthropology in particular – deal with the most unpredictable of all beings: an individual. Indeed, economics takes its roots from social sciences or, to be more precise, from social philosophy. Some currents within economics can still be counted among the social sciences. Over the course of time, however, economics strongly tended to be perceived as closer to a natural science than to an allegedly unscientific social science. This was because economics had a strong claim to describe the world with mathematical precision and to uncover social laws that would be like the laws of nature discovered by sciences like physics (Knight 1964 [1921]: 3–5; Von Mises 1996 [1949]: 7–8).

This difference is reflected also in the distinction between *descriptive* and *prescriptive* (otherwise known as *normative*) approaches to human behavior. Whereas anthropology and other social sciences employ the former, economics usually follows the latter. The reason for this is that economics – especially its theoretical branch – is interested in constructing models. To be sure, this draws upon people's factual behavior, but it prefers to deal "with tendencies, i.e. with what *would* happen under simplified conditions never realized, but always more or less closely approached in practice" (Knight 1964 [1921]: 5). It does not focus on individuals because models have to be built out of a much bigger sample. Economics as a part of praxeology also tries to establish what means should be applied in order to achieve the chosen ends, that is, how making right decisions could be facilitated (cf. von Mises 1996 [1949]: 10). Anthropology, in turn, deals with actual human behaviors. It investigates not which decisions are right or wrong in achieving goals but rather the cultural context of decision-making and of the evaluations of people's actions (cf. Boholm et al. 2013: 98).

The differences between the two approaches can be inferred from their different assumptions on human rationality, if not human nature as such. Although beginning from the late 18th century economists did not focus on the individual, they presented a strong conviction about what an individual is like and they tried to reduce every economic theory to individual preferences and choices. The reduction consisted in cutting economy off from other spheres of human life and cutting an individual off from his or her social context. Thus the construct of *homo economicus* was created, a solitary individual who aims at maximizing his or her personal advantage. Later on, this view of human behavior was transformed – by John Stuart Mill and representatives of the Austrian School among others – into such theoretical approaches as *methodological individualism, rational choice theory*, and *maximization of expected utility*. All these approaches are based on two complementary pillars of economic rationality: instrumentality and philosophical egoism. According to them, an individual desires to maximize his or her utilities and as a rule strives for financial profit. An individual has to assess the values of his or her goals and calculate the risks attached to these goals. Eventually, the individual will choose the goal connected with the greatest utility – i.e. with the relatively highest value and the smallest risk – and act in order to achieve it. In this perspective, relations with other individuals are needed only when these others are useful. Thus, relations constitute means or instruments to achieve particular goals. This is because human beings are egoistic by nature (Bruni 2002: 43–51; Hann and Hart 2011: 172–173; cf. Zinn 2008: 5).

In this view, preferring to drive a car over travelling by air is irrational and biased towards the individual's beliefs. Driving a car takes more time, is more dangerous, and costs more money if one has to travel long distances. The same applies to the well-known example of the lottery, as described by Adam Smith: people are more likely to make small stakes in the hope of gaining big prizes – even though the probability of winning is close to zero – than to risk the loss of bigger sums, even when the chance of winning a prize is much higher. Smith did not approve playing the lottery at all, but he pointed out that once somebody decided to play, it would be rational for them to do so when the chances for enriching oneself were real (Smith 2007 [1776]: 70).

Drawing upon actual behaviors such as driving a car and playing the lottery, the representatives of the social sciences criticized the approach of economics to human behavior as falling into traps of reification, dehumanization, and isolation. The first error lies in treating decisions as objects and not as social constructs. The second one pertains to disregarding factors that shape decision-making other than pure rational calculation – emotions, the decision-maker's past experience, personality, and so on. Finally, the last error is that economics tears a particular decision out of the entirety of an actor's behavior, thus ignoring the processual nature of making decisions. Economic anthropologists, in their turn, try to uncover different rationalities and to investigate the context and values that human behavior is entangled in. Rather than judging some actions as not fitting the superior rationalistic model, they seek for justification of such actions in the internal logics that depend on particular situations. Even when people perfectly know what lies in their self-interest, they sometimes do not follow it due to their regard for certain values or under the influence of emotions, choosing to act e.g. in more altruistic manner (cf. Boholm et al. 2013: 99–101; Hann and Hart 2011: 7–9, Krzyworzeka 2013: 130; Salecl 2010: 7–8). In other words, what anthropologists point out is that the case of Robinson Crusoe is absolutely exceptional. In most situations our lives consist of, there is always the native Friday. It is the relational dimension that governs our rationalities and economic behaviors.

Among the broad theoretical achievements of economic anthropology, my perspective can be placed within the institutional approach of economy inasmuch as I investigate how economic behavior is embedded into the broader cultural and historical context of a society (Granovetter 1973; Granovetter 1985; Polanyi 1957). I find this approach particularly useful to study markets – open-air markets to be more precise. Somewhat paradoxically, this approach enables me to show how the operation of the markets and their actors are shaped by non-market institutions, ranging from the household to ethnicity, religion, and government. In other words, following Karl Polanyi (1957) I understand economy as a instituted process and I show how market behavior is to a considerable extent embedded in non-market factors.

❖ Looking at the Russian transformation through Aristotelian lens

In looking at the Russian transformation I treat the household – or family – as the basic unit of analysis. I argue that for many traders, the will to sustain one's household constituted – and still constitutes – the basic incentive of their activity. The meaningful notion of *vyzhivaniye* (*survival*), which will be broadly discussed in this study, has a collective dimension. That is, one survives together, both owing to and for the sake of one's closest family and friends. My inspiration to adopt this approach stems from Aristotle's writings on ethics and politics, in which he provides a theory of economics as an art.

Etymologically, *economy* derives from *household* (Gr. *oikos*). *Economics*, at its roots, is the science of *household management* (Gr. *oikonomia*). In Aristotelian terms, a *householder* is a person who is responsible for all other members of the household. This means not merely providing them with goods necessary to live but also caring for their health, moral upbringing, and creation of all the conditions necessary to make their happiness possible. Members of one's household – and this can be extended further to other relatives and friends – constitute for an individual his or her *koinonia*, i.e. the primary community (Aristotle 1998; Aristotle 2011). It is the Aristotelian householder on whom I base my view on entrepreneurship. Such entrepreneurship is rooted in different communities, on one hand placing restraints on an individual, but on the other hand providing them with certainty about material life (cf. Gudeman 2010: 214, Hann and Hart 2011: 18–20).

Managing the household is something that in practice is an ongoing daily activity. It consists of countless small decisions: what to buy, what to use today, how to organize one's time, what values to be guided by, and so on. In difficult times, it demands a lot of effort and inventiveness, but it is always an absolute priority. After all, you have to secure the existence of yourself and your family first and only then can you undertake any other plans you might have. Thus, maintaining the household forms the basic level of the economy, an economy that can be dubbed the "subsistence" or "everyday economy". At the same time, it constitutes a prototype of entrepreneurship and all other economic behavior. Entrepreneurs, according to this definition, are people resourceful enough to sustain their household. Contrary to Aristotle, who saw the man in this position, in the reality of the Russian transformation it was often women who played this role. In the 1990s, sustaining a household meant for a vast number of people an everyday struggle to make the ends meet and to make tomorrow possible. This has continued to be the case for many people today.

I follow Aristotelian ideas also in order to understand who were producers of a specific economic culture that appeared in 1990s Russia. Culture abhors a vacuum and there has been no cultural vacuum after the twilight of socialism in Central and Eastern Europe. What took place was a rapid social change which consisted of active encounters between East and West, of constructing the new out of the rubble of the old, and of using past experience in the present reality. It is, however, important to see emerging capitalisms in the region not as simple applications of Western economic and cultural patterns but as their local and creative appropriations (cf. Kovacs and Zentai 2013: 3–7). One has to see – and this is to supplement the claim concerning the role of traders in Russian transformation – local actors as agents who contributed to creating new economic cultures. This study explores one such culture, the economic culture of Russian petty traders.

3. From informal Soviet economy to Russian petty trade

The 1990s in Russia saw an enormous explosion of petty trade. This fact was, somewhat paradoxically, both surprising and completely understandable. In general, the last time when activities connected with small entrepreneurship were legal in Russia was the late 1920s. After Joseph Stalin broke with the New Economic Policy (NEP) in 1929, trade carried out by private individuals – with the exclusion of a few narrow spheres – became strictly forbidden. Thus, it could hardly be said that the masses of people who engaged in trade in the 1990s followed any pre-revolutionary trading tradition. Therefore, the question which appears here is about the background of the aforementioned eruption. One answer could be that petty trade exploded out of nothing. To buy and to sell things is a very simple activity and one does not have to inherit it from previous generations. However, supporters of another answer – including myself – maintain that very few things spring up *ex nihilo*. To trade in socks might not be a particularly sophisticated matter, but to make it one's livelihood is, especially when it is connected with distant trips to get the goods.

The claim I put forward in this chapter is that petty trade in Russia in the 1990s derived from various practices developed during Soviet times. I track down its origins in the role of the so-called informal economy in the socialist economic system and the conditions of everyday life in real socialism. On the one hand, one could accuse me here of stating a truism. There is nothing more obvious than the claim that a present fact derives from the past. After all, an effect must have a cause. On the other hand, however, the claim is paradoxical: how could a centrally-planned system which was said to support the citizens' passivity and collectivity become a breeding ground for individual entrepreneurship? This question is of central importance to me.

In other words, the question raised above is about continuity. Systemic transformations in Central-Eastern Europe are a story of discontinuities and ruptures. At the same time, however, one can see some connection between the past and the present. Some thinkers (Sztompka 2000; Tischner 1992) claim that this continuity lies in the lingering of the so-called *homo sovieticus*, a breed of man inherited from the previous epoch. According to this view, the Soviet man is first of all a passive being, a client of the state that taught him to obediently carry out orders. Now the *homo sovieticus* expects from the capitalist state the same privileges he got from the socialist one. I, in turn, claim that the continuity between the past and the present lies in similar human behaviors, in parallel ways of solving basic life problems. Investigating the phenomenon called by Alexei Yurchak *entrepreneurial governmentality* (2002), I focus on people's everyday experience rather than on the relations between an individual and the state. There is a continuity of memory and experience in the totally new conditions that emerged after the collapse of socialism. However, the process of transforming what the former Soviet citizen knew and adapting it to the new reality was long and painful. It might even still be in progress.

As late as the late 1980s, practically nobody involved predicted the fall of socialism in Central and Eastern Europe. The disintegration of the Eastern Bloc and the dissolution of the Soviet Union shocked both the oppositionists in the East and the whole army of Kremlinologists in the West. The thorough consideration of the revolutionary events under Mikhail Gorbachev's rule between 1985 and 1991 was done *a posteriori*, when scholars developed "the hindsight necessary for understanding what they failed to grasp before" (Verdery 1996: 19). As is well known, the Owl of Minerva only takes flight at dusk.

Since the events surprised even their main actors, their consequences were not foreseen either. In the 1990s, according to various estimates, tens of millions of people were engaged in international shuttle trade (cf. Kapralova and Karaseva 2005: 402; Mel'nichenko 1997: online). Such a massive appearance of petty trade was not included in Yegor Gaidar's plan for restructuring the Soviet economy, for the transition from socialism to capitalism. That is why the described eruption can be counted among the *unpredicted effects* of the Russian systemic transformation (Ivleva 2010: 9). At the same time, however, in the face of the economic catastrophe of the early 1990s the outburst of petty trade was not only understandable but necessary as well.

Yurchak writes about the dismantling of the Soviet Union: "Although the system's collapse had been unimaginable before it began, it appeared unsurprising when it happened" (2005: 1). The aim of this chapter is to understand the analogical paradox, to grasp how people who had worked in the socialist state enterprises all their life one day found themselves ready to become private entrepreneurs.

Researching the socialist informal economy

In the 1970s, the notion of the informal economy was associated with Third World countries only. It was in a paper based on ethnographic fieldwork in Accra, Ghana that Keith Hart (1973) proposed the term. Hart argued that masses of poor city-dwellers, while being officially unemployed, could hardly be described as non-working. On the contrary, crowds of people engaged in informal activities, ranging from market gardening to petty trade to hustling and pickpocketing, either trying to make a living in ways not included in national statistics or combining such activities with official employment (Hart 1973: 69–70). Later on, one could notice, the term became a buzzword and started to be commonly applied, regardless of the political and economic order in question (cf. Hart 2001: 146; Hart 2011: 145–147).

Here the problem begins, as there is a lot of confusion about its range and even how one should call it. As one of the authors points out, this other type of economy has been described as underground, black, irregular, shadow, unofficial, hidden, unrecorded, parallel, informal, submerged, clandestine, unobserved, concealed, dual, cash, twilight, moonlight, second, counter, back-door, and subterranean (Gutmann 1985: 26). Among Russian researchers perhaps the most popular term is *informal economy* (*neformal'naya ekonomika*), though the notion of *shadow economy* (*tenevaya ekonomika*) is also in use. Apart from that, scholars often describe the informal economy in the Soviet Union as the *second* one (*vtoraya*) (Klyamkin and Timofeyev 2000; Timofeyev 2008; Barsukova 2004; 2009).

These terms point out various aspects of the informal economy. What all the scholars agree on is that the informal economy includes economic behaviors not controlled by the state and not accounted for in official statistics. Such a definition, however, is too wide. Perhaps, the fact that one picks mushrooms in a forest and eats them at home is not of great economic significance, but if one exchanges them for berries with a neighbor or sells them by the road, then from an economic perspective something more significant happens. These activities are not registered and are informal.

However, they differ greatly from black market weapons trade. What connects all the aforementioned terms is their reference to the formal, official, white, on the ground, etc. economy – the economy whose legality and visibility nobody doubts.

With regard to the countries of the Eastern Bloc, the parallel economy could be viewed as a practical consequence of the centrally-planned system. First of all, however, it should be stressed that the centrally-planned economy and the market economy were both economic, though in different senses. The strategy of the socialist economy was generally not oriented towards gaining profit. In the socialist economy the term *profit* – absolutely central to the market economy – did not play an essential role. The prices of goods in the East were much more disentangled from their cost of production than in the West (cf. Humphrey 2002b: 58; Verdery 1996: 26). The goal of the socialist enterprise was primarily about fulfilling the plan, forming a work collective, and providing the people with the means necessary to secure a certain standard of living (cf. Clarke 1999: 55–61).

As it turned out, the ideas that drove the socialist economy led to the constant growth of the informal economy within it. The cause of this process lied in an inner contradiction: the socialist economy was not able to produce enough resources to accomplish its own plans. That is why János Kornai (1980) listed shortage as among the features of the socialist economy. The central plans remained more or less a fiction, because one simply cannot possess, or at least no actual central planners have possessed, sufficient knowledge to plan the production and the consumption of everything in a country. The particular plans and the raw materials granted to execute them were always a matter of negotiations between managers of state enterprises and planners in the center. What took place in fact was a kind of "planning tender" (Clarke 1999: 74; Kozłowski 2004: 51–54) in which the goal of a manager was to receive the biggest possible input while being obliged to produce as small an output as possible. The more managers padded their needs, the more inaccurate was the planners' data and the more likely that bottlenecks would appear, which caused a further deepening of shortages. As a result, central planning was something "which the center neither adequately planned nor controlled" (Verdery 1996: 20).

The surplus a manager would receive in the process of bargaining with officials could be exchanged with other managers for needed resources that were unavailable otherwise. The barter exchanges that took place between the enterprises outside the plan – but were done in order to fulfill it – constituted an important part of the socialist second economy. Of course, the officials on every level of the economic and political hierarchy knew about the necessity of barter exchange between the enterprises. They silently approved of it, and thus somehow legalized it, because it would bring them some additional personal profits.[7]

The second economy in socialism was thus the practical consequence of the formal one, because in the face of constant shortage of goods it was impossible to execute

---

7 For further literature about the informal economy and the barter exchange between state enterprises in socialist and 1990s Russia, see: Burawoy 2001; Humphrey 2002b; Katsenelinboigen 1977; Kornai 1980; Sampson 1987.
With regard to relationships between managers of state enterprises and public officials, it must be noted that the tradition of "feeding the official" in CEE is much older than socialism and it successfully survived its collapse (cf. Chwalba 1995; Klyamkin and Timofeyev 2000).

the central plan in an official and fully legal way. The informal economy petrified the political economy of socialism and for decades enabled its functioning while eventually contributing to its fall. As far as the party officials and the managers of state enterprises gained personal profits from the functioning of the second turnover, they were not interested in changing the *status quo*. This strategy had its limits, however, because at a certain stage the socialist economy crossed the threshold of efficiency. State resources, like any other, were not inexhaustible. Starting from the late 1970s, the economies of the countries of the Socialist Bloc plunged into deeper and deeper crisis.

❖ An anthropological approach to the informal socialist economy

One can describe the informal socialist economy from what one might call an economic point of view, i.e. how this economy functioned as a whole, but an anthropological approach is also possible here. Namely, it is the position of an individual in such a reality which is worth a deeper look. The economy is understood here more as an individual strategy of behavior than as the economic system of a state. Of course, in practice the two perspectives are difficult to separate, as the example of state managers showed. Nevertheless, in the approach I call here anthropological, the informal economy is strongly tied to the work by which one earns his or her living, whether it be one's main occupation or an additional one (cf. Sowa 1990: 25).

Moreover, I follow what Vadim Radayev in his typology of informal economy defined as *gray markets*, that is, producing legal goods using illegal means such as tax evasion. *Gray markets* differ from the *black* one which cover producing illegal goods using illegal means such as drugs (Radaev 1999: 38). What interests me here are *gray* rather than *black* activities, i.e. deeds that are illegal but justified from a non-juridical perspective. Linking all these aspects offers a possibility of narrowing down the notion of *informal economy* at the individual level. Namely, I define it as the activities that use illegal means to achieve legal goals and that are usually tied to one's work. Another crucial element here is the aspect of the non-juridical justification of the action.

I will present here three examples of informal practices associated with work: two from literature and one from my fieldwork diary.

1) "An employee took a desk from a state-owned factory, intending to resell it. He left the desk in a truck near his apartment building until it could be delivered to the intended purchasers. But to his dismay, it disappeared. He complained bitterly to his neighbors that *people are dishonest and immoral*" (Wedel 1986: 15).
2) Jerzy Chłopecki, in his study of the Polish state-run workplace writes about a warehouseman in a factory. The man was an "unbelievably efficient tradesman" and "very generous distributor" at the same time. The thing was, if you wanted to take out something bigger from the factory, you had to arrange it with him. For instance, if you meant to "privatize" a drilling machine, he could legalize it by writing that the tool was broken and had to be thrown away. You could also bring an old drill from home and exchange it for a new one. Of course, the warehouseman did not do it all for free; he wanted to profit from his illegal deeds in spite of his responsibilities. "Sometimes I work at night, writing up protocols and calculations, which you then have to get signed by finding somebody drunk or by standing him a bottle of vodka," he said, explaining how he arranged with his superiors to keep the deficiencies in the warehouse from becoming a scandal (Chłopecki 1990: 138).

3) In the early 1980s in Krasnoyarsk, Konstantin (B2) led an informal construction brigade working as *shabashniki*.[8] One of their commissions was to build houses for the forest enterprise (*lespromkhoz*) close to the city. They got the job because Konstantin knew people from the head of the company. On paper there were 20 people employed in the construction but Konstantin carried out the work with just four other men. They worked hard from dawn till dusk for several weeks. When Konstantin came back home, his wife Yuliya (B3) asked him "Did you rob a bank?" because he had earned 5,000 rubles. The average salary then was around 250 rubles and a brand new Zhiguli automobile cost 4,000 rubles at that time.

The first example shows the difference between private and public property in socialism. In Russian popular usage the terms *vynesti* (to take out) and *ukrast'* (to steal) had completely different meanings, although the first was in fact an euphemism for the second.[9] The desk was taken out, not stolen. Taking something out of one's workplace is positive, as it means that one is smart and capable of taking care of his or her relatives. A popular saying went "He who does not steal from his workplace, steals from his own family." A similar meaning – among many others – was also attached to the verb *organizovat'* (to organize). If you took a bucket of paint from your workplace to paint your fence, you did not steal it, you *organized* it. The desk was stolen only when it disappeared from outside the employee's apartment building. At that time, the item was for the man and his neighbors something private, belonging not to nobody (the state) but to somebody. Social acceptance of theft from the workplace was in stark contrast with the point of view of state law, for stealing public property was punished more severely in socialist countries.

The second example also deals with the specific attitude to state property. Here, however, the story involves not stealing a single thing, but working on one's own account while serving as a state employee. What the warehouseman does is to "privatize" items and one might call him a private manager of state property. We can only suppose how much additional profit – beside his salary – he would gain. It should be noted, however, that this profit was certainly not only in cash or perhaps not even mainly in cash. When he needed to arrange something with his superiors, he did not bribe them with money, he stood them a bottle of vodka.

---

8 There are various expressions in Russian denoting informal practices attached to work. One can use such terms as *rabotat' na levo* – to earn money on the [left] side, *levachit'* – to illegally use work time, tools, or products of public labor for one's profit, and *shabashka* – moonlighting (*shabashniki* – moonlighters; *shabashnichat'* or *shabashit'* – to moonlight) (Kuznetsov 1998: online; cf. Yurchak 2002: 282, 318). Sergey Ozhegov's dictionary lists a second meaning of the term *shabashka* as well: a small thing that one takes from work, previously a small piece of wood taken home by a carpenter (Ozhegov 1949–1992: online). Personally, I have never encountered such a meaning. Also, hardly anyone recognizes today that the term comes from the word *shabash* (*the Sabbath*). Thus *shabashka* would originally denote working on holy days.

9 The example from Janine Wedel's book concerned socialist Poland. Here, however, I explain it in Russian terms to avoid introducing here another language, one irrelevant for this study. Firstly, the Polish terms are in this case very similar to the Russian one. Secondly, the example describes an attitude to public property, an attitude which was quite widespread in the former Eastern Bloc.

However, bribery might have been the case as well. As many as 70% of the managers of state enterprises interviewed by Chłopecki would offer a bribe "for the good of the enterprise" (Chłopecki 1990: 195). One should nonetheless bear in mind that money is a very delicate thing and bribery is somewhat awkward between people who remain in a deeper personal relationship. The warehouseman did his friends from work favors rather than selling something to them. Analogically, these favors were then exchanged for other favors, be they services or consumer goods. The warehouseman could have been perceived as a decent man, independently of his second activity at work or even precisely because of it (cf. Nowak 1979: 169). Here, again, the juridical assessment of a deed differs greatly from the perspective of social morality. The same applied to corruption of officials by managers of state enterprises. It was understood that the good of the enterprise requires such actions and there is no other way to arrange certain things.

The third example illustrates the contribution one puts to work done on one's own. It was almost impossible not to be officially employed in the Soviet Union. An unemployed healthy male person was a *parasite* and as such was deprived of many civil rights, sometimes including even possessing a personal identity document. Thus, to work several weeks on their own, Konstantin and his friends needed either to be formally employed elsewhere or to have protection from somebody in law enforcement. Both entailed having connections. Connections were also necessary for Konstantin to get the job. That points to an important feature of the informal economy, namely that it is highly personalized.

It is interesting to see *shabashniki* from the perspective of the phenomenon Kornai (1980) called *unemployment in the workplace.* In a system where everybody is employed, there is a need for high-quality workers.[10] If there are no good workers to hire, you have to employ two bad ones to do the same job. Another aspect of this phenomenon is that socialist enterprises usually had overemployment not only for ideological reasons but also due to an essential characteristic of the socialist economy, namely shortages. When raw materials required for production are not delivered, workers have nothing to do, but when the materials eventually come, enterprises have to work twice the usual speed in order to fulfill the plan. Nonetheless, when there is nothing to do in the workplace and the work is not well paid, the discipline of workers falls. Thus, people began to use the workplace for their own ends, the more so because moonlighting could bring very high profits as the case of Konstantin shows. Moreover, it seems that both the employers of the moonlighters and the receivers of their services were glad, all because of the speed and cost of their work.

Putting it in Max Weber's (2001 [1905]) terms, formality and informality differ in the degree of rationalization of working conditions. This pertains to issues such as regular working hours, life insurance or pension contributions. Granted, Konstantin earned a lot from his *shabashka* activities, but due to the informality of his work he was deprived of any social services. Furthermore, as I have indicated in my definition of informal economy, informality is closely linked with illegality. Thus, informal doings often expose an individual to criminal factors. As a matter of fact, Konstantin himself

---

10 In the big cities of Soviet Russia there was a constant lack of free hands for work. Since the beginning of the 1960s *shabashniki* would come to central Russia from the Caucasus. They were employed on contracts for specific tasks, with omission of the registration requirement (cf. Marciniak 2001: 340–341).

experienced both being detained by the police and being forced to cope with bandits. In general, one can note that informal economic activities are much more connected with incalculable aspects of reality, such as informal rules, personal skills, semi-legal or illegal phenomena. Thus, they are much more prone to uncertainty than the formal and regular economic behavior (cf. Hart 2011: 142, 145).

The everyday world of real socialism: Aleksandr Zinovyev and his *homo sovieticus*

The term *real socialism* was introduced by Soviet propaganda in the 1970s in order to explain deficiencies in fulfilling socialist promises. It was supposed to justify the existence of certain relics from the previous era such as secret police, money, or censorship. The state itself was – according to communistic ideology – meant to vanish in the course of building socialism and, in consequence, of communism. In fact, however, the state was omnipresent. Nikita Khrushchev promised that communism was to be achieved in 1980. Leonid Brezhnev, in turn, announced in the Preamble to the 1977 constitution the realization of socialism (Constitution 1977: online). The next Soviet leader, Yury Andropov, stated in 1983 that the Soviet Union had entered the stage of *mature* (*zrely*) or *developed* (*razvitoy*) socialism and that this stage would be long-lasting. Public opinion understood the statements well: the socialism that Soviet citizens had was the real one, which would not improve and communism would never come (cf. Sarnov 2002: 472–474).

The first writings of Aleksandr Zinovyev coincided with Brezhnev's announcement of achieving socialism. As a matter of fact, however, they can be viewed as a counterpoint to the image of socialism provided by propaganda. In this section, I would like to refer to Zinovyev's thought, because I find it particularly useful in the context of the claim that I have put forth in this chapter.

It was Zinovyev who coined the term *homo sovieticus*, the Soviet man. In 1982 he published a novel under such a title (Zinoviev 1982). Despite his philosophical background, Zinovyev sometimes declared himself a sociologist; understanding the essence of communism and of communist society was the main goal of his writing. His perspective is attractive to me for two reasons. Firstly, he described socialist society from the perspective of the everyday life of so-called ordinary people. This life, in Zinovyev's opinion, could be considered only in connection with the natural environment of the Soviet man, i.e. the conditions of living and working in the Soviet society. Secondly, the image of *homo sovieticus* worked out by Zinovyev is much more nuanced and differs substantially from the picture that was produced later from neoliberal positions (Tischner 1992, Sztompka 2000; cf. Buchowski 2006; Kovács and Zentai 2012).

I shall follow an example in order to show a specific dimension of living in real socialism, namely the problems an individual experienced in acquiring basic consumer goods. In Zinovyev's novel *The Radiant Future*, the narrator describes the success of buying toilet paper. His family, full of joy, comes back home swathed in bundles of toilet rolls: "Be that as it may, we're sure of clean bottoms for the next six months. I have to admit that in this respect (that of bulk buying of things that are hard to come by) Tamurka is a master. When all Moscow is on the hunt for bicarbonate of soda or washing powder, we almost certainly have plenty in reserve. She even manages to keep two months' supply of onions. This toilet-paper triumph brought a measure of unity and harmony to our disintegrating, although still healthy and Soviet, family" (Zinoviev 1980: 115).

For public opinion in the countries of the Eastern Bloc, queuing for toilet paper took on an almost symbolic character and, together with memories of empty shelves in shops, appears very often in accounts of the socialist era. It constituted perhaps the most visible proof of the inefficiency of the socialist economy. For Soviet citizens it seemed paradoxical that their state was able to explore outer space but it had certain problems in supplying its inhabitants with a sufficient amount of toilet paper. The source of such paradoxes was the fact that the socialist state supported production but ignored consumption. Furthermore, it was primarily heavy industry production that was supported.

Because of the permanent shortages of consumer goods and services under socialism, people often had to find novel ways to obtain them. In order to do so, they developed a repertoire of strategies that ranged from legal and quasi-legal to openly illegal. In Zinovyev's view, socialist society is an arena where everybody fights against everybody. According to communist propaganda, in capitalist societies *man is a wolf to man*. Zinovyev's diagnosis of socialist societies sounded very similar. Here, *man is a comrade wolf to man* (Zinoviev 1981: 140).[11]

Zinovyev revoked the Marxian slogans that were supposed to manage distribution of goods in future regimes. In socialism the rule will be: "From each according to his ability, to each according to his work". In communism, in turn, it will be quite similar: "From each according to his ability, to each according to his needs" (cf. Marx 1875: online). Both rules were included in Soviet constitutions, namely Stalin's from 1936 as well as Brezhnev's from 1977 (cf.: Constitution 1936: online; Constitution 1977: online). Zinovyev confirmed that, indeed, in the Soviet Union socialism had been achieved. However, this socialism was not that dreamed about by Marx but the real one. As a consequence, the rule of distribution of goods in real socialism – or communism, since Zinovyev used both terms synonymously – changed into: "From each according to his ability, to each according to his social position" (cf. Zinoviev 1981: 141–142).

Ordinary Soviet citizens had to fight for basic goods and services that in theory they were entitled to. In practice, however, they needed to get them by bribes or by prolonged queuing. Above all, the *homo sovieticus* fought for a higher social position. In the Soviet reality it was not the money one earns which is the most important. One person could buy particular things that another could not, although they both had the money. Most crucial was the position one had in the social hierarchy, as it was one's workplace that allowed one to buy in special shops goods that were unavailable to others. The majority of people had to wait for hours on end in queues for toilet paper or sausage, while others could get these goods without any problems. To the man in the street it was difficult to buy medicine he needed, arrange a place in kindergarten for his child, or receive a referral for vacation. "In all his or her vitally important problems, a citizen must overcome what at first seems like absurd problems, which

11 Zinovyev's diagnosis of Soviet society reminds one of the Hobbesian state of nature. However, the conditions of life in real socialism come from the character of power, not from lack of it as it was put by Thomas Hobbes (Hobbes 1991 [1651]: 89). According to the Soviet philosopher, the communist power took from the society all restraints that guard Western civilization such as property rights, religion, moral restrictions, or obligations under the law (cf. Zinoviev 1981: 140–142; Zinoviev 1994: 61–64).

end up costing them their time, energy, health, and resources" (Zinoviev 1981: 49). The family from *The Radiant Future* managed to buy toilet paper because it was smart enough to get information about delivery quickly and to get a good place in line for it – in other words, it fought against other families for it. The narrator's family got toilet paper, while other families did not.

In general, Zinovyev's *homo sovieticus* constituted a by-product of utopian social engineering projects that were propagated in the USSR in the 1920s and the 1930s. The Soviet man was the opposite of such figures as Pavel Korchagin or Alexei Stakhanov. He was not a stalwart supporter of the cause, rather he was demoralized, lazy, and egoistic. That does not mean, however, that he was deprived of any positive features. As a matter of fact, Zinovyev criticized socialistic conditions of living and the people shaped by these conditions only while socialism persisted. After it was gone, the philosopher, in the face of dramatic changes in Russia, began to praise and to miss both the Soviet Union and the *homo sovieticus.*

What Zinovyev emphasized in his analyses was the constant effort an individual had to put into securing ordinary hand-to-mouth existence. Acquiring basic goods and services in real socialism cost a human being a great deal of time and energy, and it required social capital. As Janine Wedel put it, in order to preserve your existence in socialist society you had to *arrange* or to *manage* things, and arranging many things was possible only through *acquaintances* and *connections* (1986). Personally, I remember from my childhood keeping a place for my parents in lines for shops. Maintaining a household was so time-consuming for them that they were unable to rely only on their own abilities. Besides, if you were with a child you could buy larger amounts of certain products. My parents cooperated and helped each other with family and friends. If there was news to share, I would run and knock on every acquainted family's door in our housing block, shouting e.g. "They dropped oranges!" (meaning the local market had a delivery). In cases more sophisticated than buying toilet paper, an individual constantly had to scheme to solve everyday problems and was often forced to act unconventionally *on the left, outside of official channels, under the table, through the back doors* (cf. Wedel 1986: 43). All that meant that people in everyday socialism had to be extremely active just to meet their basic life needs.

This activeness, I argue, is in contradiction to another view on *homo sovieticus,* proclaimed by thinkers much more critical of communism than Zinovyev or Wedel. Such Polish thinkers as Józef Tischner and Piotr Sztompka blamed some negative aspects of systemic transformation precisely on people burdened with what they see as the *heritage of homo sovieticus.* For Tischner the *homo sovieticus* constituted a post-communist form of *escape from freedom* – it was a being who did not accept responsibility for his or her own existence (Tischner 1992: 125; cf. Fromm 1968). Sztompka, in turn, maintained that people who had problems in adapting themselves to the new reality in Central-Eastern Europe were immersed in – what he called – *a culture of the Communist Bloc.* Because of this, people made unfounded demands on the social security system, requesting their workplaces to be safe, asking for minimal social security benefits in the fields of healthcare, the pension system, and education, they demanded places in kindergartens for their kids, and so on. The mistake they made came from being indoctrinated during the socialist period, and lay in the conviction that they deserved all these entitlements unconditionally. As the culture of the Bloc brought no ways to cope with problems that emerged after the fall of socialism, these people were facing unexpected changes which made them frustrated and helpless (Sztompka 2000: 73–74).

Contrary to this view, I argue in this book that masses of people in Central-Eastern Europe, and particularly in Russia, found themselves ready to face the rapid social change. This was so exactly because of the experience they brought from everyday life in real socialism. The experience of informal labor activities and of maintaining a household – i.e. being an Aristotelian householder, as I saw it in Chapter 2 – required entrepreneurial skills and enabled people to adapt themselves to extremely difficult conditions of living in 1990s Russia.

The kolkhoz market and *barakholka*: petty trade in socialism

Petty trade during socialism should be seen in the set of individual informal practices tied to one's work. The possibilities of engaging in private trade within the Eastern Bloc varied from country to country. Whereas in countries like Poland and Hungary a significant private sector – including small craftsman entrepreneurship – existed legally, in other countries – Romania and the USSR, for instance – it was much more limited. As in other parts of the informal economy, the activities in petty trade spanned from legal to completely illicit strategies. In general, private trade was perceived as an activity that was at least suspicious, if not criminal. With Stalin's ending of the era of the NEP, to buy something at one place and sell it at a higher price in another, where there was greater demand, was declared an act of speculation. The trader speculates because he or she does not add a value of labor to the product but only grows wealthy on other people's work.

In the USSR legal forms of private petty trade consisted, first of all, of kolkhoz markets, where peasants sold their produce. They were allowed to trade, because theoretically they sold what they had produced with their own hands – what they got from the kolkhoz or what they grew on their plots. Besides, as the Soviet planned economy was unable to meet consumers' needs for foodstuffs, especially in the cities, the kolkhoz markets played an important role in food supply. Although the prices at these markets were much higher than in state shops, urban residents gladly bought from there, because the products were always available and of better quality.

The second type of private petty trade permitted by the Soviet state was the flea market or the second-hand clothes market, the so-called *barakholka*.[12] Both trading at the kolkhoz market and at the *barakholka* was taxed. Another aspect that links the two places of commerce is that only owners of the products were allowed to sell there – in the case of the *barakholka* they could only sell old, second-hand merchandise. Although both types of markets were legal, they were perceived by the authorities as a kind of challenge to the socialized retail sector, a challenge that was linked to the possibility of excessive private profit and the appearance of criminal elements. That is why these markets were tolerated relatively rarely, and there were much fewer flea markets than the kolkhoz ones. In Krasnoyarsk – which in the late Soviet era had almost one million inhabitants – there was only one *barakholka*, which as a matter of fact was located outside the city limits. In Moscow the last large flea market in the

12 According to Kuznetsov's dictionary a *barakholka* is a place where old, used clothes (*barakhlo*) are sold directly (Kuznetsov 1998: online).

district of Perovo was banned in the mid-1950s. In Kiev, in turn, the last flea market was shut down in 1975 (Katsenelinboigen 1977; Sik and Wallace 1999).

One could say that the communist authorities were quite right about petty trade. In practice, the two aforementioned types of markets tended to become centers of semi-legal and illegal transactions. This was especially the case of trade in exotic fruits and flowers which operated under the aegis of kolkhoz markets and was highly profitable. In Moscow, but in other cities as well, these branches were practically monopolized by peoples from the Caucasus, namely Georgians and Azerbaijanis. It was almost exclusively women who stood behind the counters, as they were less prone to the accusation of being a *parasite*. Traders in both fruit and flowers tried to get around various legal regulations. For instance, they sold produce of good quality *under the counter* for higher prices than the upper price limits set by the government. They also moved outside the kolkhoz markets into more attractive locations and they extended official working hours. This required some informal arrangements with local police officers and other authorities. In general, the flower and exotic fruit branch of trade can be called semi-legal, as the activity and produce itself were legal, but various practices associated with it were less so (Katsenelinboigen 1977: 71–72).

Barakholkas, in turn, stocked not only second-hands clothes but also commodities produced informally by local craftsmen or brought from abroad. There were a number of consumer goods plagued by shortage in the Soviet Union, including male and female clothing, shoes, furniture, cosmetics, coffee, chocolate, and so on. Some things could be sent in packages from the outside of the USSR, delivered by people from abroad – there were for instance famous "drinking boats" from Helsinki to Leningrad – or by individuals who came to the USSR with various delegations. There were also special categories of Soviet citizens who on different occasions could travel abroad, such as sailors, athletes, artists, scientists, and party officials. Whereas for Soviet citizens bringing foreign currency and private turnover of it was strictly forbidden, the ban on conveying goods was only loosely obeyed. One could even bring a jacket bought in the West to a special state-run commission shop, where the price would be individually set. This, however, was unattractive because it required personal registration and bringing bigger amounts of things would be suspicious. Therefore, it was more advantageous to sell it to a speculator even at a lower price but leaving to him or her the risky task of finding a buyer. The speculators worked at the *barakholkas* – some of which, like the ones in Odessa and Tallinn, were well-known and supplied other regions of the country – or in private apartments in the cities where there was no flea market, like Moscow (Katsenelinboigen 1977: 77–81).[13]

Dealers in clothes were called *fartsovshchiki*[14] and they were particularly prosecuted by the law-enforcement bodies because of their frequent contacts with foreigners.

---

13 Katsenelinboigen (1977) classifies markets in the USSR according to the extent of legality and color-codes the classes: with *red* denoting state network shops; *pink* – state commission shops; *white* – kolkhoz markets and *barakholkas*; *gray* – the market of renting apartments and dachas, private services performed by craftsmen, the market of flowers; *brown* – informally imported deficit goods, the market of spare parts; *black* – illegal goods, foreign currencies, gold, drugs, prostitution.

14 Nikolai Komlev's dictionary of foreign words and phrases lists the following as possible roots of the term *fartsovshchik*: *force* (French and English); *forsch* (German – daring, bold), *forsować* (Polish – to force, to insist, to strain; *forsa* – colloquial term for money);

Among such traders was Konstantin (B2), the *shabashnik.* Admittedly, he did not buy clothes from foreigners but he used to sell the female clothing made at home by his wife at a *barakholka* near Krasnoyarsk. He told me he had been detained by the police many times, although after receiving a bribe the officers would always let him go and he could carry on trading. The bribes constituted an additional cost for Konstantin, but in spite of this, the business was very profitable.

Indeed, what is striking about the goods plagued by shortage, whether informally imported or produced locally, are the prices. In 1950s Warsaw a lipstick brought from the West cost 175 zlotys and was offered by a private seller, while one could eat a proper lunch for a mere 10 zlotys. A good monthly salary was 800 zlotys at that time (Tyrmand 2009: 271). One kilo of oranges in 1980s Poland was worth 1/10 of the average monthly salary and a sheepskin coat could be obtained only for as much as 15 months' pay (Wedel 1986: 19, 145). In the mid-1970s in the USSR a woman had to work for almost a month to afford a pair of fashionable shoes, as they cost 70 rubles and the female monthly wage was around 90. For a male synthetic suit a private trader would charge 120 rubles, about 10 times more than it cost in the West (Katsenelinboigen 1977: 78, 85).

The explanation for such extremely high prices is ultimately the demand. If a trader is able to sell oranges for 1/10 of the average salary, then this means that someone is ready to buy them at that price. A pair of blue jeans could cost several months' pay and jeans might be ideologically suspicious, but there were always individuals eager to have them. This phenomenon demonstrates, above all, that in societies in which everybody was equal, some people were more equal than others: senior party officials, diplomats, renowned artists, high-ranking military men, senior managers and engineers, etc. They were even millionaires in the socialist countries. In Soviet Georgia in the 1970s Otari Lozishvili, the manager of a few state factories, earned an unbelievable 4 million US dollars per year (Marciniak 2001: 346). After all, the western lipstick, the oranges and the blue jeans constituted luxury goods in real socialism and as such were markers of social status. However, people from lower social strata would also strive to acquire such goods. This was often linked to moonlighting and all the additional earnings that sometimes significantly exceeded one's monthly salary, but some categories of people could even earn larger amounts of money officially, such as those working in the Soviet Far North. Such people often would find themselves in the situation of having cash but not having anywhere to spend it. When they had the opportunity to come to the big city, they had an easier access to the second market.

The shortages of consumer goods were not only engendered by the structure of the planned economy itself but were also artificially caused. If, in general, furniture was a commodity associated with shortage in the USSR, for some people it paid to hamper the production of it in state factories in order to produce it in the very same factories unofficially. In other words, state enterprises manufactured underground goods that were then traded via informal channels. The people engaged in such activity were called *tsekhoviki* as the branches of this production were *tsekhi* (workshops, departments of production, guild). Lozishvili's fortune came from possession of several *tsekhi,* which means he somehow owned the state factories themselves. The partners

---

for sell (English). The dictionary goes on to explain that *fartsovshchik* in the USSR meant a person buying currency and valuables from foreigners in order to resell them at a high profit (Komlev 2006: online).

of his business were almost all members of the Central Committee of the Georgian Communist Party (Marciniak 2001: 345–346). However, the socialist system of redistribution underwent privatization on the level of individuals as well. A seller in a state shop could, for instance, *put aside* a hard-to-get piece of furniture for an acquaintance and sell it for a price higher than the official price.

It is very difficult here to distinguish *gray* activities from *black* ones. There were different degrees of illegality – selling flowers was semi-legal and selling jeans illegal, but more legal than selling foreign currency, for example. Social acceptance of the activities might have aligned with this scale: flowers are needed more than jeans. Nevertheless, if somebody made too much money too easily, people would start to hold him in contempt even if they needed what he traded. And, ultimately, trading in blue jeans might be illegal but the person involved in it does not harm anybody. That differentiates such a speculator from a millionaire who cheats large groups of people, and thus is a criminal.

Mikhail Gorbachev and perestroika

Mikhail Gorbachev became the General Secretary of the Communist Party of the Soviet Union in March 1985. Just after taking the position he announced the necessity to overcome *zastoy*, the economic stagnation of the Brezhnev era. To this end, he proposed a program of *uskoreniye* (acceleration) of the state economy, a new plan that was supposed to modernize the country. In January 1987 he declared the beginning of further, broad systemic reforms that went down in history under the slogans of *perestroika* (restructuring) and *glasnost'* (openness).

Some scholars describe Gorbachev as a pure ideologist, perhaps the last man who believed in communism (cf. Aslund 1995: 28; Morrison 1991: 35). He himself expressed his commitment to communist ideas on multiple occasions. For instance, while explaining the goals of the restructuring of state economy and politics, he depicted perestroika as a new revolution. The need for a new revolution could be perceived at that time as blasphemy inasmuch as the October Revolution was supposed to be the last one in history. Therefore, Gorbachev modestly admitted that he did not dare to equate perestroika with the October Revolution, but he still used the term *revolution* and presented himself as a follower of Vladimir Lenin (Gorbachev 1988: 49–50). As a matter of fact, his unique and ultimate goal was to preserve and to reinforce the Soviet Union (Aslund 1995: 27). The way to achieve it was not by introducing capitalism but improving socialism according to the rule *more socialism and more democracy*.

However, on the way to better socialism, and attempting to solve the catastrophic economic situation in the country, Gorbachev and his aides took a number of contradictory decisions. Between 1985 and 1991 the Soviet leader announced over 10 radical plans for improvement of the economy, but they were never implemented. As one scholar put it, "Perestroika failed in large part because it was not tried" (Boettke 1993: 2). Gorbachev imagined something impossible: he wanted to create an ideal communist system that, nevertheless, would include some capitalist solutions. He intended to combine communism with elements of a free market and the Communist Party's monopoly on power with a multiparty system. He strived to liberalize censorship of the press but was shocked when journalists started to criticize him. His attack on alcohol probably had lofty aims but it brought the state budget dozens of billions of rubles in losses. Furthermore, it contributed to sugar shortages as people started to produce moonshine on a large scale (cf. Aslund 1991: 78–80; Brooks 1988: 5; Dobrovolsky 2010: online; Pikhoya and Timofeev 1999: 341). He proclaimed openness and at the same time gave his silent approval when national demonstrations in Vilnius, Riga and Tbilisi were bloodily put down by the army. Thus, in general, "Gorbachev's project was doomed by its own internal contradictions" (Gallopin 2009: 20).

But even if perestroika had been implemented, it would not have revived the economic system. It is questionable whether this system with problems that had accumulated for decades could ever be healed. However, what was most fallacious was the very idea of solving problems of a very general nature by old means: pumping huge amounts of money into heavy industry could not fix an economy that demanded consumer goods, innovation, and free competition. Funding the idea was possible only by taking out loans from abroad and printing massive amounts of money. Such moves always result in a rise in inflation.

The prices in the Soviet Union, however, were controlled by the government. This caused large disparity of cash and cashless transactions between enterprises and the state. In 1986 as much as 14% of the state budget was allocated in order to cover the difference between actual food prices and the significantly higher costs of food production (Brooks 1988: 6). In the 1990s such disparities led to enormous debt of state enterprises which needed to be subsidized from the budget. Inflation was also a time bomb; it was present, although stifled by price limits, revealing itself in grave shortages of state-produced goods and rapid rises in the prices of privately traded goods (Aslund 1995: 49). When the price limits were removed one by one in 1991 and 1992, inflation skyrocketed.

"A dilettante, a helpless theoretician, which makes him a chronic bungler, a poor practitioner, and an unprincipled politician," such are the words with which Gorbachev has been described by Russian historians (Pikhoya and Timofeev 1999: 349). They also added such expressions as *treacherous* and *conceited.* Zinovyev, instead of using the term *perestroika,* proposed a new one: *katastroika* (Zinoviev 1990).[15] For him, not only Gorbachev but also Yeltsin were criminals that collaborated with the West in order to betray Russia (Zinoviev 2005: 51). The last Party Secretary's mistake was that he confused his noble plans with real possibilities to fulfill them. "The basis for Gorbachev's efforts was his belief that Soviet communism was reformable. This turned out to be wrong; but if he had not believed that reform was possible he would not have been able to destroy Soviet communism" (Aslund 1995: 28).

My interlocutors also held negative opinions about Gorbachev. They usually saw in him the main culprit of the systemic change. It had been his reforms that later resulted in the closing of factories and, in general, the collapse of Russian industry. When I asked Lila (T51), who had been working at different markets in Krasnoyarsk for 20 years, why she had engaged in trade, her immediate answer was that it was Gorbachev who made her and countless other people traders. As she complained, it was Gorbachev who destroyed the USSR and began to privatize state enterprises. The television factory where Lila once worked was turned into a shopping center and this very fact she found difficult to bear.

---

15 A neologism combining two Russian words *katastrofa* (*catastrophe*) and *stroyka* (*construction*). It is the title of Zinoviev's 1990 novel, which takes place in the fictional city of Partgrad (Party City). The novel, in general, criticizes the realities of perestroika.

Critical statements toward Gorbachev are common among Russian citizens. It suffices to point to the share that he received in the 1996 presidential elections: 0.5% of votes.

❖ Wavering between contradictions: state and new forms of entrepreneurship

One of the most discernible contradictions in the era of perestroika was the Soviet policy towards private and cooperative entrepreneurship. The Brezhnev Constitution allowed some forms of individual labor activity (Constitution 1977: online, articles 13 and 17). Private entrepreneurship was legal only as a part-time job performed in one's free time; one had to be employed in a state institution first. In addition, the law forbade hiring external workers, which meant only members of one family could work in an entrepreneurial activity. The law was aimed first of all at agriculture production, the growing number of small private pieces of land, as well as some spheres of construction works and traditional handicraft. In practice it was very difficult to obtain permission for other types of activities, as the constitution gave more freedom than the actual executive legislation that regulated the issue. All the legal restrictions sharply limited number of people wanting to start a private activity. In 1986 only 97,000 entrepreneurs were registered in the USSR (Aslund 1991: 154).

The Soviet public opinion was familiar with examples from other socialist countries where private and cooperative entrepreneurship functioned on a much bigger scale. In Hungary, Poland, the German Democratic Republic, and China one was allowed to run one's own company full-time and to hire employees. Examples from East Germany, where the private sector successfully supplemented the drawbacks of the state one, particularly excited people's imagination. That is why reformists within the Communist Party began to prepare new regulations on individual economic activity. Just before the new legislation came into life, however, the bureaucratic machine was launched in the completely opposite direction.

As was discussed in depth above, the range of the informal economy within Soviet socialism was colossal. At the 27th Congress of the Communist Party in March 1986, Gorbachev announced "the struggle against unearned income". *Netrudovye dokhody* – unearned income – was a vague term, not defined clearly in the Soviet law, although present in the 1977 Constitution (Constitution 1977: online, article 13). In practice, it meant every economic activity not linked with one's work within socialist structures and not controlled by the state. As far as the word *netrudovye* translates as *non-labor*, the perverse denotation of that expression was that if you did not work for the state you did not work at all, so you were not entitled to an income. Every income, especially significant ones, could therefore be regarded as illicit. At the Congress Gorbachev talked about overuse of state property and functions, condemning for instance the practice of bribery among Party officials. However, the acts that were issued by executive bodies in May 1986 specified these statements in a particular way: the further down the body was in the legal hierarchy, the sharper the regulations concerning the private sector were.

What in Gorbachev's plans was a part of the operation called *glasnost* and was intended to diminish corruption among Party members, in practice turned into action against ordinary citizens. The scale and the style of measures taken reminded them of Stalinist times. Various symptoms of *private-property psychology* were persecuted:

informal earnings in public retail shops, unofficial housing rents, use of public transport for private gains. The strictest restrictions hit the kolkhoz markets.

Under the pretext of fighting *the middleman*, various law-enforcement bodies heavily fined the kolkhoz traders and confiscated their goods. Another target of persecution were deliverymen: in the USSR transport for private goods was hardly available and the vast majority of such services were provided by *levaki*, state-employed drivers performing additional jobs on the side. Labor on private plots and selling surpluses at the market was not only legal but even officially supported by the Soviet authorities. Nonetheless, in 1986 the facts diverged greatly from what was being proclaimed by propaganda. Even absolutely legal produce was seized by the militia and destroyed on the spot. The immediate result were much bigger shortages and higher prices. *Is a cucumber guilty?* read the title of an article published in a local newspaper in Perm, where the prices of vegetables jumped three times. In Volgograd oblast hundreds of greenhouses were razed to the ground. Similar to the Moldovan vineyards destroyed within the anti-alcoholic campaign,[16] now it was the tomato that became the enemy of the state. Bans on the sale of different goods spread indiscriminately over the country, e.g. flowers were banned in Leningrad, and deliveries between regions were seriously limited (Aslund 1991: 158–162). All this was happening in a country where on vast territories it is difficult or simply impossible to grow even potatoes.

In the second half of 1987 the struggle against illicit incomes was softened as it had failed to reveal corruption among Party members and to tame the informal economy. Another issue was that a new law concerning private labor came into effect. To make a long story short, the Soviet leaders intended to stimulate the activities they had just stopped persecuting.

The *Law on Individual Labor Activity* that came into effect in May 1987 was the first law since the times of the NEP that extensively dealt with private entrepreneurship. However, the possibilities opened up by the legislation were very limited. The Law restricted these activities only to part-time and family businesses. Only some consumer services and traditional handicrafts were allowed, while many others such as photocopying, production of jewelry, or perfumes remained forbidden. The goal of the legislator was not to let the private sector compete with the state. It was, however, not the limited fields of running a business that made life particularly difficult for potential entrepreneurs, but the taxes and licenses that one had to obtain. Income taxes for private enterprises reached 65%, whereas for natural persons they did not surpass 20%. Licenses, in turn, were issued by local authorities for periods not exceeding 5 years, so no permanence was guaranteed. In addition, the procedure of getting such a permit was long and many of the applications were rejected. All those factors made the gateway into the private sector very narrow (Aslund 1991: 164–167).

❖ The Law on Cooperatives and the story of Konstantin and Yuliya

Of much greater importance was the *Law on Cooperatives* that came into force at the beginning of July 1988. For Soviet communists the hardest nut to crack was private

16 In 1985 during the struggle of the anti-alcoholic campaign around 80,000 hectares of vineyards were annihilated. Valentin Bodyul, who was the head-manager of the famous Moldavian winery Cricova at that time, said that workers of the factory were forced to chop grapevines down during the weekends (Dobrovolsky 2010: online).

ownership, and that was the reason they looked on private enterprises with such suspicion. Cooperatives, in turn, could be explained in terms of socialist cooperation and as such were much easier to accept. But while the existing cooperatives, such as kolkhoz farms, did not in fact differ much from ordinary state enterprises, the cooperatives of the new type lived up to their name. They were self-managed, self-financing, owned by their members and market, and they were profit-oriented. According to the Law, establishing a cooperative required only 3 people and no special license was required. The companies could engage in every legal activity, including trade, and they were granted with wide prerogatives – their members could work full-time there and had the same rights as state employees. The taxes set for them were low: private taxi drivers, for instance, had to pay 560 rubles per year, but those who belonged to a cooperative paid only 30, although there was no real difference between them (Aslund 1991: 167–171).

Konstantin (B2), the *shabashnik*, with a few companions, founded a cooperative as early as 1988. He claimed that it was the second cooperative registered in Krasnoyarsk and the only one that has survived since the 1980s to the present day. They dealt with a job they knew very well, namely construction and renovation. The cooperative grew very quickly, and one year after its launch almost 500 people worked there. Whereas in state factories people earned 250–300 rubles per month, in Konstantin's cooperative even an ordinary workers received 3,000. This was possible because the quality regime in the company was strictly obeyed. Since customers paid good money for the commissions, all tasks were done quickly and well. In addition, Konstantin introduced a *capitalist* or *normal* – as he called it – system of management into the cooperative.

> *"In a socialist enterprise which employed, say, 100 people, there were 60 workers, the rest were various managers, presidents and inspectors. Which is to say, those 60 guys had to support 40 freeloaders. In our cooperative, out of 300 employees there were up to twelve managers and foremen, the rest were workers who received real pay for their work."*[17] [B2].

Konstantin's cooperative worked, albeit not without its share of problems. The company created competition for the socialist sector and deprived it of its best workers, who abandoned their workplaces and moved to the enterprise for its high salaries. For these reasons some officials, pressed by managers of state enterprises, wanted to close down the cooperative. Konstantin, however, had good connections in the circles of the First Secretary of the city's branch of the Communist Party. The second aspect was that his company was a very specific one and as such was respected even by the officials who were unfriendly to him.

Namely, Konstantin, his wife Yuliya (B3), and their business partners organized something that could be called a charity foundation. Every employee of the cooperative was offered the possibility to give up 10% of his or her earnings and hand it over to

17 Because of the conditions in which I did my fieldwork I have not recorded the majority of the interviews I conducted. After a talk I wrote down all I managed to remember, doing my best to note down the literal words of the interview partner(s). I use just italics to provide a quotation when I am referring to a recorded interview, so the quotation is exact (albeit translated, of course). If, in turn, I use italics with quotation marks, I am referring to an unrecorded talk, which I am recalling from my field notes, so the quotation is not exact.

the fund. The employees were completely free to make the donation but – surprising as that may sound – everybody agreed. Their incentives – as Yuliya, who was responsible for the fund, explained to me – were of a social character. In the dark times of the late 1980s, they felt that when they earned so much, they had to share with people in need. Yuliya used the fund to buy, for instance, toys for orphanages, sponsor meals in kindergartens and schools, or pay nurses to stay longer with patients in hospitals. In other words, the fund supported the state in the duties it was unable to fulfill. That was one of the reasons, as Konstantin and Yuliya consider, that officials did not cause trouble for the cooperative.

In practice, however, it was not that easy even to open a cooperative. Local authorities often prohibited such initiatives, or even used illegal means to stop them. Public opinion expressed its opposition towards it as well. A few reasons for such reaction were already mentioned in the case of Konstantin: the competition they posed for state enterprises, the outflow of workers, and the high earnings. Prices of goods and services delivered by cooperative enterprises were also criticized, as they were 3–4 times higher. One can understand this, however, if one takes into account the better quality, the low competition and, last but not least, the immense shortages. Private enterprises could impose high prices because people would buy their products anyway.

A passage between epochs: the late Soviet entrepreneurship

Studies showed that labor efficiency was 3–10 times higher in cooperatives than in the state sector and that was a positive phenomenon. At the same time, however, salaries there could exceed state salaries by even dozens of times. This stood in glaring contrast with the notion of socialist equality. The Soviet press raised the alarm that new millionaires were going to appear. The examples were as follows: a sovkhoz tractor driver earns sometimes only 2–3 rubles per day, while a cooperative worker makes up to 100 rubles by producing bra fasteners. Members of one of the cooperatives in the Donbass received up to 5,000 rubles per month by producing plastic jewelry and haberdashery accessories. One has to compare these figures with the highest Soviet official payment at that time: 1,200 rubles earned by people working on the government level. Interestingly enough, the amount of private cooperatives and their acceptance varied significantly depending on the region. In the Baltic republics there where up to 20 times more services covered by such enterprises than in the Siberian and Central-Asian regions (Aslund 1991: 172–176). Konstantin's cooperative was perhaps one of the few in the entire Krasnoyarsk Krai.

In general, the case of the Russian private sector in the late 1980s allows one to see several maladies of the transitory period. The legal regulations were unclear and the state economy unstable. Elements of free market and private entrepreneurship were introduced too quickly and without changing the whole system. This situation brought many problems.

As was stated above by discussing the roles of state managers, in a chronic shortage economy it is the supplies that constitute the main difficulty. As a matter of fact, no Soviet enterprise – be it a state or private one – could function without purchases or barter exchanges in an informal market. Such operations were often linked with bribes. Thus, Soviet leaders intending to develop private activity did not give entrepreneurs the possibility to run their businesses legally. Private entrepreneurship would work if the free market were there, but it was not. As a consequence, many entrepreneurs were prosecuted. As one of the authors sadly notes, these people who were called *speculators* where the one who did not get political permission for gaining big profits. One person could have serious problems for possessing 100 US dollars, while another person could run completely freely a business worth millions. In the early 1990s there were as many as 127,000 entrepreneurs imprisoned in Russia. Most of them were sentenced during the reign of the *liberal* Gorbachev (Marciniak 2001: 343). Such persecution perfectly fit the Soviet dialectic of encouraging and deterring private labor activity. On the one hand, it was difficult for an entrepreneur to enter the official market and they had reasons to fear persecution, on the other hand high business profits made overstepping the law a tempting opportunity.

Besides this, entrepreneurship in the USSR was attractive for organized crime. It was in the Brezhnev era that gangsters started to infiltrate various spheres of the shadow economy, as it was more profitable and less risky for them than openly illegal activities (cf. Humphrey 2002b: 111). In 1979 in Kislovodsk a special secret summit took place, where *tsekhoviki* agreed to pay criminals (so called *vory v zakone*, thieves-in-law) 10% of their profits for protection, for not kidnapping their children, and for their own safety (Gurov 1990: online). Following this example of the southern part of the country where *tsekhi* were particularly developed, similar agreements were reached in other Soviet regions, too. According to one estimate, by the decline of the USSR 75% of cooperatives in Moscow and as much as 90% in Leningrad made such payments (Humphrey 2002b: 109). Nevertheless, *vory* extorted money also from other categories of people who got "unearned income": medical doctors performing private practice, managers of restaurants and, of course, traders. One survey from the late 1980s showed for instance that each and every tradesperson at the Rizhski Rynok in Moscow – now the biggest flower market in the city – paid 25 rubles daily for protection. There was not a single exception (Gurov 1990: online; Schwirtz 2008: online).

In spite of the criminal phenomena, in a country where there was virtually no private entrepreneurship before, the new regulations presented an important turn. Although the *Law on Individual Labor Activity* opened up limited opportunities and had a small impact on systemic changes, the *Law on Cooperatives* can be seen as one of the most essential reforms introduced by Gorbachev. Cooperatives became a channel of using the experience and skills people had acquired in their work and everyday life. In 1990 private cooperative enterprises hired already 5 million employees (Aslund 1991: 178). Even for those who did not engage in such enterprises personally, such companies constituted a point of reference. It was during perestroika when the new class of Russian entrepreneurs was born.

It was also the time when social classes began to reshape themselves. The result of that process depended basically on the starting point and possibilities. If one looks at many of the former or current Russian oligarchs – such as Mikhail Khodorkovsky, Boris Berezovsky, Vladimir Potanin, Mikhail Prokhorov, or Vagit Alekperov – it turns out that they were Komsomol or Party members and they started their business careers in the late 1980s. Participation in communist structures gave an individual the necessary contacts and opened opportunities for engagement in the most profitable fields, such as international trade or the mining industry.

In general, the new entrepreneurs often flourished because of their network of acquaintances. Alena Ledeneva made *blat* – a notion stemming from prison slang and denoting informal connections – a key concept of the Soviet and Russian economy, both the shadow and the official one. In her approach, the economy was based on favors and the favors themselves – as far as they entail mutuality – served as the first measure to build social and business ties (Ledeneva 1998). The socialist experience turned out to be very useful here, as "when the Soviet Union fell apart, this system of favors was employed to cope with a variety of new problems" (Gallopin 2009: 2).

During perestroika Party members started to secure privileged positions in the society, exactly as though they felt the country would not last long. One factor that contributed to the fiasco of perestroika, and is usually mentioned as one of its primary causes, was the resistance of bureaucracy to reforms. After the first restructuring plans failed, Gorbachev understood that economic reforms could not be carried out without political ones. These, in turn, were very difficult to implement, since Party officials wanted everything but openness.

Their resistance was linked to the proliferation of the informal or shadow economy. In the course of the 1970s and 1980s a market which can be called an administrative one appeared in the Soviet Union. In the last decade of the USSR, only a roughly estimated one third of the GDP was distributed according to central planning. The rest was divided in the administrative market, where state goods and services were privatized by Party officials. Each good and each decision could be an object of bargaining between different interest groups (Marciniak 2001: 88). Rather than limiting or legalizing different fields of the informal economy, Gorbachev's reforms only supported their expansion.

By the time that the reformers realized the enormous size of the shadow economy and corruption, it was actually too late to fix anything (Pikhoya and Timofeev 1999: 340). In the last days of the USSR "bureaucratic competition substituted for economic competition" (Boettke 1993: 5). Zinovyev would probably put it differently – everybody took as much advantage of public goods as only his or her social position allowed. One can also note that Party officials became the most dexterous entrepreneurs (Stoner-Weiss 2006: 31). Perhaps the best description for that practice was given by the political scientist Leonid Radzikhovsky, the headline of whose newspaper article read "The nomenklatura have exchanged *Capital* for capital" (1995). Communist Party officials displayed a lot of energy in grabbing up state property. Eventually, the very existence of the USSR was put up for auction. The communist nomenklatura privatized the state itself and this was a nail in the Soviet Union's coffin (Marciniak 2001: 89).

Certainly dismantling the Soviet Union was not the goal of Mikhail Gorbachev: "Gorbachev did not become General Secretary to reign over the demise of the Soviet empire" (Boettke 1993: 3). Rather, he imagined himself as a new Stalin; a powerful communist leader who would drag his homeland out of the deepest crisis. Perestroika in such a scenario would correspond to the NEP, a policy that, incidentally, ended with a unprecedented wave of terror (cf. Zinoviev 1988: 36–58). Nevertheless, the processes of perestroika and glasnost launched by Gorbachev are largely seen as the main catalyst leading to the subsequent dissolution of the Soviet Union in December 1991.[18]

---

18 One can actually see that a whole set of different factors contributed to the dismantling of Soviet socialism. Although before 1991 hardly no one expected the USSR to fall, after the fact a very rich literature has been produced concerning the political, economic, and social changes that lead to the dissolution and the consequences it had for 1990s Russia. For historical studies see for instance Afanasyev 2001; MacKenzie and Curran 1999; Marciniak 2001; Owen 1995; Pikhoya and Timofeyev 1999; Sakwa 1999; for studies that place an emphasis on social and economic change, see for instance Aslund (1991;

Why did we liberalize import? Simply to fill the shop shelves. It was as simple as that. To avoid hunger. The state had no money to buy the products itself – how on earth did those ants [petty-traders and/or smugglers] that instantly started to crawl around manage to do their thing? I was shocked. Seventy years of up-rooting the entrepreneur spirit by all means possible – you give them freedom and boom: the next day it all just works. I was the one enforcing those changes and even I was shocked.

Andrey Nechayev (2011: online).[19]

4. Coping with uncertainty: surviving the Russian transformation

They work in order to survive, I work in order to live is what I heard during the very first interview I carried out in Russia.[20] These were the words my interview partner, Galina (T1), used to talk about labor migrants who traded at the same big open-air market. What she meant by the difference between survival (*vyzhivaniye*) and life (*zhizn'*) was that migrants worked all day long without any rest. They sent to their families in Tajikistan, Azerbaijan, and other former Soviet republics everything they had managed to save. They tried to save money on everything they could, whether accommodation, food, transport, etc. Thus, they were surviving. In contrast, Galina *enjoyed her life*. She was Russian and a local. She lived with her family in their own apartment, and she could afford fresh meat every day and designer clothes for her children. Thus, *to survive* means in this context to be able to fulfill basic needs of one's family at the expense of hard efforts. *To live*, in turn, denotes an existence which is secure and has a certain standard. Galina worked to live, while the migrants lived to work.

The aim of this book in general, and this chapter in particular, is to look at the Russian systemic transformation from the grassroots perspective of the people who were forced to survive it. Essentially, it is the dissolution of the Soviet Union and the civilizational catastrophe which hit the whole post-Soviet region afterwards that made masses of people dispossessed, that is, deprived of their life savings, jobs, places of living, social status, and networks (cf. Humphrey 2002b: 21ff.; Nazpary 2002: 13–16). Several questions stem from this fact: how do individual people cope with rapid social change? How do different individuals handle basic economic uncertainty, i.e. phenomena that appear at the national and supranational levels, such as instability of currency, unemployment, or poverty? How does one live in order to survive?

1995; 2002; 2007a; 2007b); Boettke 1993; Lane 2002; for anthropological studies of late Soviet society see for instance Boym 1994; Ries 1997; Yurchak 2006.

19 Nechayev was a member of Gaidar's Government: from November 1991 until February 1992. He served as the Deputy Minister of Economy and from February 1992 until March 1993 he was the Minister of Economy.

20 The Russian sentence was: *oni rabotayut, chtoby vyzhit', ya, chtoby – zhit'*. Lexically, the difference between the Russian verbs *zhit'* (*to live*) and *vyzhit'* (*to survive,*) is slight. They usually occur in similar contexts, but sometimes – as this sentence shows – they can also be juxtaposed.

For many dispossessed people in the entire postsocialist space, trade became the sole possibility of survival. Through trade many of them managed to adjust themselves to a completely new reality, a reality which was constantly changing (cf. Mandel and Humphrey 2002). However, my first claim – which overlaps the first main claim of this study – is that it was also these people who made the change. Obviously enough, as I mentioned in the Introduction, various global processes connected with the dissolution of the Soviet Union were not started by the Russian citizens who engaged in trade in the 1990s. The vendors had to cope with changes that were imposed on them from above. At the same time, however, I argue that they actively shaped the course of the Russian transformation. As I will show, they managed not only to stabilize their economic situation but also – as they were so many and thus so potent – to positively influence the condition of the national economy. It is thus also the interrelations between macro- and microprocesses that this chapter aims to point out.

The second claim I put forward in this chapter is that the activity of a petty trader takes place in a field of necessity and freedom. Survival belongs to the order of necessity. *They have to survive*, Galina told me about the migrants. When I asked people *Why did you engage in trade?*, the answer I repeatedly received was *Life forced me to.* I also often heard that *one has to live* or *you must feed your children, after all.*[21] The opposite pole to the necessities of life, in its turn, is constituted by freedom, which I equate here with a space of possibilities for entrepreneurship.

As I wrote earlier, I understand entrepreneurship in basically two senses. The narrow one concerns business activity and coping with a peculiar tangle of various risks and uncertainties characteristic to this activity. In the unstable political and economic situation of post-Soviet Russia, running one's own individual enterprise, whether international shuttle trade or something else, required from an entrepreneur a great deal of innovativeness and diligence in dealing with necessities. In the wide sense, entrepreneurship points out a resourcefulness – people's skills to secure the livelihood of themselves and their families. If one has to engage in trade to feed one's children or if migrants send home all the money they save, this goes to show that traders do not operate just for themselves. Thus, they should be seen in relation to their basic communities, their families and closest friends. On the one hand, traders work for the sake of their communities, but on the other the communities constitute the most important resource for them. An entrepreneur in this view is a Aristotelian householder who both contributes to and profits from his or her household.

In this chapter, I reconstruct the reality in which the massive eruption of petty trade took place. Consequently, the chapter is devoted mainly to a description of the

21 Galina said *im nado vyzhivat'*. The two other sentences quoted above are *a zhit'-to nado* (one has to live) and *a detey zhe kormit'-to nado* (you must feed your children, after all). There is a problem in translating the Russian word *nado*. You can interpret it as *have to* or *must* but actually it means both. This problem, however, is tiny if you compare it to the challenge of translating the central notions of this book: *zhizn'*(life) and *vyzhivanie* (survival). Because of the irreducible existential context of these notions I find them untranslatable into English. Thus, what I try to do is to reconstruct the cultural context in which these notions are used.

1990s. Grasping the changes of the 1990s is crucial to understand not only the current status of petty trade but also the whole socio-political situation of contemporary Russia. I begin with an outline of structural reforms carried out by Yegor Gaidar's government in the early 1990s and their influence on Russian society. I then move on to the concrete problems people were forced to cope with: unemployment or working without being paid, the instability of the ruble and the necessity to provide one's household with basic foodstuffs. The following parts describe the phenomenon of *chelnoki*, the shuttle traders, who in the 1990s massively travelled long distances to get goods from abroad. At the end I focus my attention on the role the issues of informality and criminality play in the activity of a petty trader. As the micro- and macrolevels inextricably intertwine, and the grassroots and top-down perspectives are inseparably entangled with each other, I constantly move between individual biographies and general social processes. What is more, I do not stop only at the 1990s but also show the evolution that trade has undergone since that time and I frequently refer to the present day.

❖ The reforms of Yegor Gaidar

In 1990 the Soviet economy was in dire straits. On the one side, state budget expenses rose because of investments in industry, thus increasing labor wages and social benefits. On the other side, however, state revenues were falling due to falling production, decreasing oil prices on international markets, and lower tax incomes. As a result, the budget deficit skyrocketed in 1991 to almost 30%. The other urgent problem was foreign debt and servicing payments tied to it. To finance his acceleration program, Gorbachev had to take out huge loans from abroad. As Anders Aslund estimates, the Soviet foreign debt rose from 14 billion dollars in 1984 to 57 billion just seven years later (1995: 47–49). However, the real amount of the debt at that time is difficult to calculate. Andrey Nechayev, for instance, gives other numbers, numbers that are – literally – incredible. According to him, by the end of 1991 the foreign debt was as high as nearly 129 billion dollars, while the state currency reserves were as low as 26 million (Nechayev 2011: online). The USSR announced its insolvency in terms of foreign loans in December 1991.

Gorbachev and other Soviet leaders were desperately seeking a solution. The so-called 500 Days Program of radical reforms, drawn up by Grigory Yavlinsky and Stanislav Shatalin, failed to gain the necessary support within the Communist Party. At the beginning of 1991 some reforms were introduced by Valentin Pavlov, then prime minister and minister of finance. The improvements were most peculiar, however, consisting basically in robbing the citizens of their money. In January, Pavlov initiated a monetary reform: the 50- and 100-ruble notes were withdrawn from circulation, with the catch that people had only three days to exchange their money and the sums of exchange were limited.

A much more serious harm to the citizens, however, was inflicted when their deposits were frozen in April 1991. The colossal amount of 600 billion rubles taken from people's savings was transferred to patch up the internal debt. In the late 1980s the Soviet Union began to suffer from a significant and rapidly growing monetary overhang. The phenomenon occurs when people have money but do not have anywhere to spend it. This was the case of the USSR, where wages were decent and grew constantly, but due to shortages people did not have possibility to spend them on goods.[22] Therefore, they kept their cash in banks. The situation was aggravated by the huge monetary emission during perestroika; too much money was in circulation and inflation was starting to spin out of control. Pavlov kept promising that deposits would be reimbursed. It was only Yegor Gaidar who by the end of 1991 revealed the brutal truth: they were gone. As the money was never given back, the result of Pavlov's decisions was that masses of Soviet citizens lost all their life savings. This was a fatal blow to their trust in the state, politicians, and the banking system.

The economic chaos was deepened by the political one. Starting with Estonia in November 1988, one after another the Soviet republics declared their sovereignty and then their independence from the USSR. The center started to lose control over some of its territories. Perhaps the strongest blow to the Soviet Empire was dealt by the Russians themselves: on June 12, 1990 the Russian Soviet Federative Socialist Republic (RSFSR) declared its sovereignty and exactly a year later Boris Yeltsin was elected its President. Since then, and especially after the failure of the August Putsch of 1991, the situation in the country was unique. One could notice a whole shadow state had emerged within the Soviet Union, with its own authorities, institutions, and internal and external policy: it was Russia.

Such was the situation in November 1991, when Yegor Gaidar became the Minister of Economy and Finance of the RSFSR. Gaidar and his team (*komanda Gaidara*) knew all too well about the necessity for reforms. The real question was how they should be carried out. The general problem was the lack of pattern – at least an intellectual one – of passing from socialism to capitalism. The goal was clear, but the way in which particular reforms should be implemented remained uncertain. A move in the opposite direction – from capitalism through socialism to communism – was thoroughly elaborated upon by Soviet science. None of the Soviet economists or social philosophers could have imagined the evolution the other way around.

It was Gaidar who was responsible for the Russian transition to a market economy, although formally he was Prime Minister only for half a year, from June to December 1992. As he claimed, at that point the state had in fact gone bankrupt and the only therapy one could use had to be a shock one (Leybin 2006: online). However, both the ideas he had for reforms and the way they were implemented were heavily criticized because of the huge social costs they brought with them (cf. Burawoy 1999; 2001; 2009; Kovács 1992; 1997; Stoner-Weiss 2006: 20–23).

Incomprehensible to mortals: how could one manage in the 1990s?

The policy of shock therapy and attempts at developing the market economy in the country resulted in economic collapse. In 1992, after the liberalization of prices, the inflation rate skyrocketed, exceeding 2500%. In the following years, inflation was actively fought by means of monetary policy, but it was only in 1996 that its rate dropped

22 As a popular joke from that time summarized it:
"- What is real socialism?
- It's this time before you can get stuff without money but when money no longer buys you anything!"
(Sarnov 2005: 473).

below 100% annually (Aslund 2002: 201). Russia's GDP decreased constantly since the dissolution of the USSR until as late as 1999, when it eventually started to grow (cf. Aslund 2007a: 139, 196).[23] The depression of the state and economy led to an expansion of the poverty sphere, the collapse of social services, and a rise in criminality.

Sharp economic decline and severe social problems hit all postsocialist states throughout Central-Eastern Europe, the Caucasus, and Central Asia in the 1990s. Economic survival became the priority for whole social classes (cf. Clarke 2002). In Russia, in the mid-1990s industrial workers went for months without being paid, while agricultural workers in many areas would not receive wages for years. The same was the case with pensioners, whose pensions were frequently suspended. There was also a significant unemployment rate. In 1863 the Russian socialist Nikolay Chernyshevsky wrote about students, "As far as his finances were concerned, Lopukhov belonged to a very small minority of medical auditors – that is, students with no government stipends – who manage to live without starving or freezing to death. Exactly how they manage, and on what means, God only knows – it's totally incomprehensible to mortals" (1989 [1863]: 91). The same question can be repeated in the case of the dispossessed people in Russia: how did those people manage and with what means?

A strategy which Russian people massively chose was engaging in petty trade. "What can I do? First I borrow. Then I stand on the street corner and sell things," says a librarian asked by Caroline Humphrey how he/she manages (2002b: 82). One must remember, however, that "selling things" can actually mean a variety of activities and in any case might be quite difficult if done in extreme economic conditions.

One of my interview partners, Kirill (V6), was already retired when we met. He used to work in Krasmash, the Krasnoyarsk machinery construction plant. It is a huge enterprise which in socialist times employed tens of thousands of workers. In Krasmash mainly weapons and space rockets were – and still are – produced, but it also made Biryusa fridges "*as a cover-up*", as Kirill put it. In the early 1990s he was fired in one of the mass layoffs. He did not change his job completely, though. Initially he started work in transport services. He had a UAZ van and whenever somebody bought a fridge in the outlet he proposed to deliver it. Later, however, he realized that a better business could be made on fridges themselves. His former colleagues, who still worked at Krasmash, sold fridges on the side, for much less than the official price. The task was therefore to intercept a customer who was willing to buy a fridge in the shop. In the garages at the very back of the factory there were alternative facilities, where you could buy the same fridge with all the necessary documents, stamps and a guarantee. Even spare parts could be acquired there and, of course, you could order a delivery. "*At the beginning we were ashamed: during socialism such activity was called speculation. However, as time went by we gained our confidence. Today if you do a thing like that, you are called not a speculator but an entrepreneur,*" said Mariya, Kirill's wife. So Kirill took on as a partner his son, Maksim, who was roughly a teenager at that time. They did relatively well together, although they had to cope with some obstacles too. Firstly, they were not the only people who sold Biryusa fridges by bypassing the company shop. As a matter of fact, there was strong competition in

23 It should be noted, however, that according to some prominent economists Russia surpassed the 1989, pre-recession level of GDP only in 2012. "For over two decades there was no improvement in living standards for most Russians," says Vladimir Popov in a conversation with Piotr Dutkiewicz (Popov 2013: 404).

that niche. Secondly, local criminals also wanted to have their share in the business, so the *entrepreneurs* had to pay protection money. And thirdly, there was the risk of bringing the wrong customer to the garages. Once Maksim saw an elegant man who got out of a good car and headed towards the shop. Maksim approached him quickly and presented his offer: the man listened to him with great interest and followed the boy to the garages. When they reached the place, to his great surprise Maksim saw that all the people who worked there were running away in panic. The secret of this strange behavior lay in the identity of the elegant man: he was the top manager of Krasmash. Everybody knew him except for Maksim, who, after all, had never worked at the factory.

Tanya (V7), today a PhD student in Barnaul, survived the worst times in the 1990s thanks to moonshine, homemade vodka. She lived with her mother and grandmother in a provincial town in Altai Krai. Her grandfather died early, and her father in turn left them when she was a baby, so the two women and one girl could count only on themselves. In order to make ends meet, they produced moonshine. What Tanya remembers from her childhood and teenage years are big cans in the kitchen, the characteristic smell, and a sense of fear. This fear was caused by two reasons: the customers, who might come and ask for a bottle in the middle of the night, and the risk of police checks. The latter had never in fact happened, as even though the neighbors knew perfectly well what was going on in the women's apartment, they never reported it to the police. Tanya was very happy when she grew up and started working, allowing the household to give up producing alcohol.

❖ Being half-free: the ambiguity of the postsocialist workplace

Darya (T65), although now a pensioner, is still an individual entrepreneur and runs her own stall at the Tsentral'ny market in Krasnoyarsk. She received higher technical education and during socialism she and her husband used to work as engineers in a cement factory. There were periods in the first half of the 1990s when they went without being paid for several months. "*It's a Russian thing*," she answered when I asked her what the reason for working was if one was not getting paid.

What Darya meant was probably the specific role a workplace played during socialism and in the 1990s. The workplace, by definition a state institution, formed the basic social unit. The socialist enterprise was not primarily about generating profit but rather about fulfilling the plan, about forming a labor collective and providing for "almost every aspect of the social existence of its employees" (Clarke 1999: 57). The enterprise would provide housing, health care, sporting and leisure activities, and so on. For a worker, the employing institution was important not only because of the salary, but also because of social services, privileges such as food cards, coupons (*talony*) to buy specific products, child care, professional training and, last but not least, the social network it provided. All this made people dependent on their workplace. These rules to some extent continued into the 1990s. This was because some of Gorbachev's and Yeltsin's policies, like the agrarian reform, were ignored throughout the country. Some collective farms in the mid-1990s still worked according to Soviet rules, running their own shops, issuing food coupons and even their own currency. Thus, employees would work there even without getting a salary because it secured for them basic goods and services (Burawoy 2001: 281–282; Humphrey 2002b: 9–13; Morrison 2007: 134–143).

That is why losing one's job was so painful in the post-Soviet reality: all of a sudden your access to many goods and services was cut off, along with your social ties. Less ties, in their turn, meant less connections and opportunities. Sometimes people were able to cope with that, as the example of Kirill showed. However, the results of unemployment were often disastrous for an individual and his or her family. In the countryside working for a collective farm might be the only job opportunity and it was not easy to become an individual entrepreneur there. Indeed, one became dispossessed, i.e. he or she was not in possession of a workplace. This was, however, a freedom that might prove difficult to bear.

It is worth mentioning that in the extremely difficult economic conditions of the 1990s the level of unemployment in Russia was relatively low and there were few demonstrations and strikes during that time. The low unemployment rate was caused by the fact that many post-Soviet state enterprises still worked according to non-profit-oriented rules, adopting social protectionism as their main priority. Even indebted and unprofitable enterprises, especially in the countryside, did not dismiss their workers. They did not dismiss them even if there was no job to do, and in fact these people did no work. The situation was also made possible by the loose monetary policy of the state: the state generously supported even completely unprofitable enterprises, because it preferred to pay the very low wages rather than unemployment benefits. The state workplace was thus treated as the major social welfare institution. As a matter of fact, however, actual unemployment rates in the 1990s were much higher than official ones (Burawoy 2001: 282; Clarke 1999: 75–76; Füllsack 2001; Wegren et al. 2003: 849–851).

"*Prishlos' vyzhivat'*" (one had to survive), said Darya when I asked her how she engaged in trade. She and her husband started to work at the Tsentral'ny in 1993. They were also still working at the cement factory. They might go unpaid for several months, but in fact they did not work for free, as after some time had passed all overdue salaries were paid. They worked in shifts of two whole days at the factory and the next two at the market. It was difficult to stand, but that was a necessity for living.

Darya's case might be called typical, since a significant part of today's vendors used to work in Soviet state enterprises and lost their jobs as a consequence of perestroika and the dissolution of the USSR. They found themselves trading at a market because they were forced to do so in order to survive. That is why some authors call them *the unwilling entrepreneurs* (*predprinimateli ponevole*) (Avilova et al. 1999: 86; Yakovlev et al. 2007). A sense of coercion was felt among tradespeople and this did not concern only the beginnings of their activity. As I heard several times for instance, *There is no other job for a woman in her 50s*. Middle-aged tradeswomen often said that within the reality of the labor market in Krasnoyarsk, they had no other choice than to work in trading until they went into retirement.

The fact of being forced to become an entrepreneur and to engage in trading stands in sharp contrast with a freedom that traders often stressed. Another woman expressed it in the following way:

> *What's good about trade is that you're a free man, you are not limited, there's no management breathing down your neck, they do not limit you. You rest when you want to, you can take leave at any time. That's what attracted me.* (T11)

Nonetheless, for many petty traders in contemporary Russian market, this freedom remains a pure possibility. If you run your own stall, you indeed do not have any boss who orders you around. However, because of many various factors – among them competition posed by other trading agents – the traders have to work as much as possible. By no means can they allow themselves to take rest whenever they want. Entrepreneurship is by its nature linked with self-discipline. Nevertheless, here the self-discipline must often be exceedingly severe. Traders, especially labor migrants, work 10–12 hours per day, 6–7 days per week. A state of permanent fatigue is characteristic for them. It is thus difficult to talk about freedom if you cannot really enjoy it. Of course, they chose to engage in trade, but the very fact that they actually did not have any other option strongly influences the self-perception of their situation (cf. Ivleva 2001; 2003; 2008). While some traders like their work, others find it to be something not for them.

Because Darya had worked in conditions that were harmful to her health, she could retire earlier. When we talked in 2012 she received a pension of 8,000 rubles (around 250 dollars) which – as she said – would not secure her livelihood. That is why she decided to continue working at the open-air market. Granted, her pension sufficed only for paying the rent for the stall, but she could earn more by trading. She traded in female clothes produced in Chinese factories in Bishkek, Kyrgyzstan. She considered Kyrgyz-manufactured clothes to be of better quality than the usual Chinese production and thus she went to Novosibirsk, 800 kilometers to the west from Krasnoyarsk, to get the goods. In Novosibirsk, the biggest city in central Siberia, there are wholesale markets, where large shipments of goods come from abroad. Darya also traded in Amway household detergents and was quite happy with that. "*There is a problem in Russia with fake products but Amway is original and of superior quality,*" she said and added contentedly that she had managed to engage a few female customers in the marketing network. Thus, Darya always combined different modes of earning her living: work at the factory, trading at the market, and receiving pension. It is here that a key to secure one's existence seems to lie. Furthermore, by engaging in the Amway network she has taken a niche which has moved her one step ahead of the competition.

However, Darya still did not perceive herself as successful or even economically secure. When I asked her about the difference between life and survival, she defined both terms clearly:

> "*Life – it would be if I worked in a factory and got a fixed salary every month. Life is something stable. On the contrary, survival is when you have to support yourself in an alternative way, it is an uncertain and a hand-to-mouth existence*" (T65).

Because of the uncertainty of profits from her trade activity and the uncertain future of the Tsentral'ny market, she saw herself in the domain of survival.

❖ Trade is a drug

There is also a force of habit in the informal trade activity and trading in general. In the Introduction I mentioned a lady (T2) who informally sold pickle cucumbers and sauerkraut by an underground station in Moscow. She came from Ryazan' Oblast. In the summertime she would grow vegetables in her garden and pick berries and mushrooms in the forest. She also bought vegetables and fruit from her neighbors and made preserves out of them. Then, in the winter she would go to Moscow and sell this produce in the street. She said that she did this in order to have money for her son's books and school supplies. On one hand, most likely she could have found another job, one much more comfortable than standing in the freezing cold and being hassled by the police. On the other hand, trading was something she learned to do and something that brought her a certain profit besides all the risks connected with her activity.

This habit to trade turns sometimes into an addiction. Tanya, for instance, said:

> *Trade is a drug. It's a drug difficult to break free from. It's a drug in that the money you collect from your merchandise you have to put into merchandise again. And again. You're simply afraid to stop because then the money will end, the money will no longer come* (T4).

Yekaterina, in turn, revealed:

> *You have to love your job! You need to love your merchandise, you have to love new merchandise! (...) I had another profession – I am a certified confectionery technologist – and I thought that would be my line of work, but there's no leaving the market! Once you grow to love it, it sucks you in and there simply is no leaving then – it sucks you in so much, it's so interesting... And the money: if you get the drift of it, if you find it in you, you will have real money. But first of all, you need self-discipline, after all you have no bosses above you. I wake up in the morning, it's cold outside, minus 15 or minus 20 degrees, but I should get up, get dressed and "forward march!" – to work* (T6).

It is paradoxical that you have to love your job, for after all, you cannot love anything or anyone compulsorily. What Yekaterina meant was that full commitment to a job can pay off in terms of financial success and personal growth. If trade is a drug, it is a very specific one. Firstly, you may be forced to take it, as in the case of Tanya, Yekaterina, and many other tradespeople who became unwilling entrepreneurs. Secondly, you may get addicted to trading, which means it provides you with a living; if you drop it, you may not survive. Unlike a usual drug, however, trading – no matter how difficult it might be – might also offer you something positive. If you start to love trading, it offers you a bit of gambling, a sense of fascination, as well as a spur for taking up risks and for the development of your business. Marx would probably call this *consciousness of necessity*: you are condemned to this freedom and it is tied to strict self-discipline, but once you choose it and make this freedom yours, you'll come to enjoy it.

❖ Coping with the ruble

One has to remember that throughout the 1990s, Russians could not trust their currency, the ruble. There was a whole Gordian knot of problems linked to the national Russian currency, and the first of them was that it was not in fact national.

After the USSR had fallen, the ruble zone did not. In 1992 most of the new countries, including Russia, introduced their own national currencies but the Soviet ruble was still in parallel use. In addition, it served as the common currency of the post-Soviet space. Only the Russian Central Bank was allowed to issue Soviet rubles, while other national banks in turn could issue non-cash obligations in that currency. And they did just that, to as great an extent as possible. Because of high inflation rates, in practice they were able to return less than they had borrowed. There was, however, a source needed for financing such operations: if somebody gains, someone else has to lose. In this case it was Russia and the Russian ruble, the currency with the strongest ties to the Soviet ruble. Russia consented to financing the debts of other post-Soviet republics due to the political reason of maintaining the unity of the Commonwealth of the

Independent States (CIS). Only in the middle of 1993 was the Soviet ruble taken out of circulation and the Russian ruble became fully national (cf. Marciniak 2004: 234–236).

The use of two parallel currencies additionally contributed to the most serious problem with the Russian ruble: hyperinflation. By the beginning of 1992 hyperinflation destroyed the working capital of enterprises and all that remained of the savings of the population. One of the consequences of hyperinflation is that too little cash money is in circulation. This factor forced state enterprises and private companies to operate almost entirely by barter. As I described it in Chapter 3, barter exchange between state enterprises was a widespread phenomenon during socialism. The mid-1990s was a period when the mechanisms of functioning barter were developed and extended to encompass an entire state. If one considers this aspect, the 1990s in Russia were a return to the past: "central planning was replaced not by a market mechanism oiled by money, but by barter transactions mediated by new forms of patronage and corruption" (Hann and Hart 2011: 131; cf. Burawoy 2001: 278–279, Clarke 2000: 178).

High inflation and its consequences haunted Russia throughout the 1990s. According to official estimates, at the end of the 1990s barter made up 47% of all economic transactions in Russia and up to 70% of transactions between state enterprises. However, in the case of the Buryat Republic this rate was as high as 80%. Local farm managers claimed not to have used cash money for five years. Entire chains of exchange were built: a collective farm, for example, would pay for electricity in meat and for gas in fruit. Obviously, the power plant and gas station had to use these products somehow, and did so by paying their own obligations with them. Still, the exchange rates for barter were set according to money prices. Networks of barter exchange were also established on an international level between Russian and Chinese enterprises (Burawoy 2001: 283–284; Humphrey 2000b; Humphrey 2002b: 86).

Russian economic history has seen the practice of establishing local currencies in times of crisis such as civil war. In the 1990s, during financial instability and frequent defaults, the practice came to be employed once again. Some big enterprises issued their own payment checks, which could be used by employees in company stores. Similar processes took place at the level of federal republics, who issued their own payment coupons that could serve as a surrogate for the ruble (Anderson 2000: 329ff.). In Moscow, tokens for the metro were used as a substitute currency. As the costs of public transport steadily grew, the tokens constituted a good investment, or rather a good form of securing one's financial means from inflation. Similarly, a 15-kopeck coin, which was used for public telephones, could – as a shortage good – be sold for 20 rubles (Lemon 1998: 22). In Sepych, a small farm town in the Urals, in the late 1990s it was moonshine that became a local system of reckoning. "Everything here is done for a bottle," said one of the interlocutors of Douglas Rogers (2005: 63).

Somewhat ironically, it can be stated that at that time a bottle of moonshine became in Sepych a good with the highest degree of liquidity. Liquidity refers most often to assets held in cash or in other transactable items that can be readily converted into cash. Cash is usually most liquid because one can exchange it for almost everything, but in a situation of hyperinflation it is better to have, for instance, a pig than the money which it is worth, because the following week this same amount of money might be able to buy you only half of the pig. Thus, moonshine was typically used to pay for small services and in everyday transactions. For example, chopping firewood for winter cost two bottles. The ruble was used to buy bigger things, e.g. a pig, but after selling the pig the inhabitants of Sepych, if they wanted to create savings, would change the rubles into dollars, which were much less prone to inflation and speculation (cf. Rogers 2005). Moonshine was commonly used as a mean of reckoning in provincial Russia.

Because of high inflation and lack of money in circulation, employees often got paid in products made by their company. This was the case especially in the provinces. The degree of liquidity of the product you got paid with was of great importance: it is easier to sell flour than an electric drill, for instance. In any case, people had to sell or exchange these products on the market. Andrey (V9) told me:

> *The entire country depended on what they call barter. Natural exchange. And literally, this was the big word of the time. You open a newspaper and there it is: barter this, barter that. But yeah, this was a huge factor. And frankly, this barter caused its share of trouble to a lot of companies: what can you live on? Let's take a toilet factory, it had no money so it paid its dues with what it made. In other words, you come back home with twenty toilet bowls as your salary. Do what you will, my friend...* (V9)

Being paid in products constitutes one of the main reasons why Russians massively engaged in petty trade in the mid-1990s. The goal of each barter chain was for an individual to find and get ahold of cash. While enterprises could operate utterly on barter, their employees could not. "Households are the primary victims of demonetization," notes one scholar (Clarke 2000: 176) – because retail trade and consumer services were performed almost entirely in and for cash. People needed money to purchase things they could not produce themselves. In this way they became traders.

❖ *Ogorod* and the dacha: to be as autarkic as it gets

Nowadays in Russia there are 50 million households and 40 million *ogorods* (*gardens*) and dachas, small plots of land found next to apartment blocks or outside the city.[24] This has significant consequences for social life. In describing this phenomenon, scholars talk about a *joined place of inhabitation* (*sovokupnoye zhilyo*) or *divided way of being* (*raspredelyonny obraz zhizni*). Besides having an apartment in the city almost everybody has a second household, auxiliary to the first. Maintaining it requires a great deal of time, effort, and money, but one also gain some profit from it (Afanasyev 2001: 17–18).

For many Russians, the dacha or *ogorod* is a way of life. It is a powerful cultural institution that often structures people's time. How people live on these plots of lands, what values they attach to them, and what social functions the plots fulfill – these issues have been described elsewhere (Caldwell 2011; Hivon 1999; Ries 2009; Zavisca 2003). In the context of this book it is enough to emphasize the key role *ogorods* played in the course of Russian systemic transformation. For masses of people this small plot of land – rows of vegetables and a few animals – formed a secure basis in a very insecure reality.

---

24 Basically, the term *dacha* refers to a plot of land with a house, outside the city, which serves predominantly for leisure. *Ogorod*, in turn, means a plot of land which is allotted for growing vegetables. Such a plot can lie in the city (e.g., by the apartment block one lives in) and has no residential structures (cf. Ozhegov 1949–1992: online). In practice, both of the terms mix, as one can have a house on one's plot and still cultivate the land.

It is first of all potatoes which secured – and often still do secure – people's existence. Andrey, who was quoted in the last subchapter, told me about the 1990s:

> *A: It was a huge help then, it was some sort of safety, something the state could not take away from you, could practically not impact in any way. You did your part, grew what you grew, and it's yours, it's on your balcony, in your fridge, in the cellar, and you can count on it. It's not a lot, it's miserable, but you're not defenseless. It was a huge help, really, and it let people survive.*
> *K: So it meant independence from the state?*
> *A: It meant independence from... Well yeah, from all accidents, from all those... It was something you secured for yourself and could be absolutely certain about.* (V9)

Thus, potatoes helped to handle uncertainty. Andrey himself came to Krasnoyarsk when he enrolled in a university in the mid-1990s. He recalled that in the first year of his studies all students were taken once to a big collective farm outside the city to help harvest potatoes. They worked for three days and in return they could take a sack of potatoes each. As it later turned out, to engage students in harvesting was by no means a form of exploitation. On the contrary, it was a form of support. When the potatoes ran out, Andrey was starving. His parents could not help him, because they were not paid and basically they were coping with the same situation. Andrey with his peers from the student dormitory then stole cabbage from the university cafeteria's warehouse. It was the only time in his life that he stole something but, as he justified himself, *really there was nothing to eat.*

For many people, growing potatoes is still of substantial importance. Potato production is laborious, tedious, and exhausting. In addition, it is often done without any support from machines. Furthermore, it is elderly, retired people who usually do most of this work, staying for months in a tiny hut and then going to the market to sell the produce. The results are unbelievable, the vast majority of Russian fruit and vegetables is grown in *ogorods.*[25]

However, Andrey mentioned that also some of his relatives – people who are quite well-off – still grow potatoes. *There is a certain inertia to it,* he said, meaning that the relatives just continue doing so by force of habit. They themselves explain it by the fact that they want to have their own potatoes, ones "not from a shop" (*nemagazinnoye*) but natural and environmentally clean. Andrey sees in this behavior also a certain fear, however. According to him, his relatives fear that something might happen and these potatoes will be needed then.

It seems that the case of Andrey's relatives is not uncommon. Many people, even from relatively prosperous families, grow their own potatoes. For a person who works full-time, this means several weekends a year spent outside the city. You have to get

25 Nancy Ries quotes President Vladimir Putin from a press conference in 2003: "In the old days, they used to allocate people six hundredths of a hectare, and then the size grew a little, and then a bit more. The strange thing is that ninety percent of the potatoes grown in the country are grown in these little private gardens. Ninety percent! And these gardens produce eighty percent of the vegetables and sixty percent of the fruit. These figures seem incredible, but these are the facts. So, I would first of all like to thank everyone out there who takes part in this work for such amazing results. And I hope that you will get satisfaction out of the work you do this season and that it will bring you good results" (Ries 2009: 202).

there somehow, you have to transport the potatoes, and so on. Thus, considering the amount of money, time, and work that potatoes require, growing them is completely uneconomical. It would be cheaper to simply buy them. The same applies probably also to many pensioners who grow potatoes, take the amount they need, and sell the rest. This is a certain economy, although it makes no sense in terms of classical economic theory. From this point of view, one could call such conduct irrational. Nonetheless, if one takes into account all crises and defaults these people went through, their behavior might prove reasonable. The *ogorod* gives them protection against markets and state policy. The *ogorod* makes one self-sufficient to a great extent and subsistence economy is as autarkic as it gets.

The *chelnoki*

The reformers from Gaidar's team quickly understood the necessity of both abolishing the state's monopoly over foreign trade and allowing it also for private individuals. It was Pyotr Aven, then Minister of Foreign Economic Relations, who prepared the new law opening up foreign trade. The key significance of this is best demonstrated by the speed with which the decision were taken. Aven was appointed to his post on November 11, 1991, and just four days later the act opening up trade was signed by Boris Yeltsin.

The reason for such a hurry was the total shortage of consumer goods. State shops were literally empty and it was not unheard of for people to line up for weeks to buy desired products (cf. Aslund 1995: 49). Many analysts claim that already in 1991 Russia was in real danger of famine. Nechayev for instance said, "When both foreign and domestic media warned Russia would not survive the winter of 1992, that we were in real danger of hunger, chaos, collapse of production, and perhaps even hunger riots – this was not a case of journalistic exaggeration, this was absolutely what the truly was" (Nechayev 2011: online). As a matter of fact, even though true famine did not appear, by the end of 1991 there were serious problems with food supply to big cities, including Saint Petersburg (Kokh 2011: online). The drastic measures taken by Gaidar helped to prevent this danger.

But who specifically managed to prevent hunger? Who saved the Russian market? Who was Russia's savior? It was the petty trader. And a specific one at that, the *chelnok* or shuttle trader.

❖ To have a flair for business: the story of Valentina and Pyotr

Valentina and Pyotr (T61), a married couple from Krasnoyarsk, started their trading trips as early as 1990.[26] Their favorite destination was Poland. In order to make their voyages more profitable, they established a tourist agency, one of the first in Krasnoyarsk. They knew how to organize the trips, because Pyotr originally came from Brest, the city on the Belarusian-Polish border, and Valentina graduated from a course in tourism in a university in Leningrad. Although the trips were called touristic, their goal was plain business. Russian traders brought to Poland tableware,

26 It proves that some people travelled abroad to get the goods even before the liberalization of trade at the end of 1991. It was, however, this liberalization that made the trips a massive phenomenon.

porcelain, pots and all which is popularly called *yerunda* (*rubbish*), i.e. needles, threads, sponges, brushes, and so on. They took also various sorts of tools, such as handsaws, vices, pliers, as well as and small electric equipment like drills or shavers. Tools were two or even three times cheaper in Russia than in Poland, so Poles eagerly bought them.

What brought the biggest profit, however, was gold and jewelry. Not only were these goods cheaper in the Soviet Union and Russia than in Poland, but gold served as a kind of currency. Shuttle traders needed foreign currency to spend abroad, but in Krasnoyarsk dollars were hardly available. At a bank you were allowed to buy only 100 dollars per person, but in any case this limitation was quite abstract, since at banks there were usually no dollars at all. To obtain gold was also not without its problems – one had to have informal connections and wait in lines to buy it – but in general it was much more acquirable. Gold was also easier to sell abroad. Thus, gold was perhaps the quickest way to earn money, buy goods, and return home. Valentina and Pyotr travelled to Żywiec, a town in southern Poland. It was a good location, because Germans would come there and buy gold for German marks, a currency much sought after in the Soviet Union.

It was in Poland that the outbreak of the August Putsch found them. The whole group of traders they led were offered political asylum with the promise of obtaining Polish citizenship. Nobody from the group accepted the offer, however. They had left their families in Russia and they wanted to be with them, no matter what the political regime in Russia would turn out to be. Afterwards, Valentina and Pyotr themselves seriously considered settling down in Poland. They were about to buy a house in Białystok, a city in northeastern Poland which they particularly liked, but eventually gave up the idea.

As organizers of trade trips Valentina and Pyotr planned everything in detail. They were responsible not only for the transport of people and goods but also for planning the itinerary, booking hotels, and arranging contacts with wholesalers in Poland. They travelled mostly by air. From Krasnoyarsk they flew to Lviv via Moscow, and from Lviv they took a bus to Przemyśl, the Polish city on the border with Ukraine. During a trip they usually visited with their "tourist" groups a few Polish cities to sell things at open-air markets there. For the money the traders earned they bought first of all clothes and shoes, but also some food and cleaning products. These goods were in great demand in the places they came from.

What the couple could not plan were borders. Once when they were crossing the Polish border, Pyotr was selected to go through customs inspection. "*I have changed my mind and I do not want to enter Poland anymore*," he said to the Polish customs officer who wanted to inspect his items. This meant that Pyotr right at the customs office decided not to visit Poland and to go back to Ukraine. The secret behind such a sudden alternation of plans lay in the goods he was carrying to Poland. Each piece of gold the officer might find on Pyotr would be confiscated as a smuggled good. Therefore, in order not to lose the gold he had, Pyotr went back to Lviv and waited for the rest of the group there. With the gold he bought there several pairs of blue jeans which he afterwards sold in Krasnoyarsk.

One of the people Valentina and Pyotr travelled with was Olga. Olga was "*a fat and impertinent woman*", as Pyotr described her when telling me her story. Because of her size, "*she took 100 grams of gold under one breast and another 100 grams under the other. Who would suspect this village woman of carrying half a kilo of gold?*".

Indeed, Olga never got checked on the border. In addition, she had good connections in Krasnoyarsk and thus was able to arrange large amounts of gold. Originally, she came from Ukraine and in the early 1990s she went back there. With the money she earned shuttle trading she bought a golden Toyota Land Cruiser. "*It was literally as if an alien spacecraft had landed*," said Pyotr, attempting to describe the impression the car made on people in Ukraine at that time. The car did not bring her luck, however. Once, while crossing a river, for some reason she drove off the bridge and into the water. This was an accident that eventually took her life.

By no means, however, was Pyotr's punch line of the story about Olga an expression of sorrow. He noted instead that Olga had only completed 8 grades of primary education for impaired children. Some people considered her mentally retarded, but in fact "*she had a real flair for business*," said Pyotr.[27] Olga must have been a true entrepreneur.

Valentina and Pyotr stopped traveling to Poland in the mid-2000s, but they still maintain their business contacts. *Polish coats* said the advertisement over their two stalls at the Tsentral'ny Market in Krasnoyarsk, the advertisement which made me strike up a conversation with them. They received catalogues from Polish clothing companies and they order wares by telephone, making their payments via money transfers. The advertisement on the stalls, however, is only partly true, as Valentina and Pyotr changed their business focus. Nowadays, it is China where they get the most of their goods and where they often travel.

The couple is one of the few traders I met that can be considered successful. Apart from the stalls, they own two other shops with jackets and coats, though these are not at another open-air market but in shopping malls. They themselves do not work full-time in the shops. They hire salespeople and replace them only when they have their days off or during special periods like the potato harvest. Pyotr eagerly showed me photos on his tablet from their holidays in Turkey and Thailand. Apparently, they stay at the market not out of pure necessity like most traders I talked to. I met them there because they make a relatively good living from it.

❖ The role of *chelnoki* in the Russian systemic transformation

Valentina and Pyotr were members of a whole army of people who engaged in small trade in the early 1990s. How big was this army? That is very difficult to assess and it would be even more complicated to calculate the size of financial turnover those people produced. The cause of these problems is, once again, the informal economy. The vast majority of deals completed by traders were not registered by the countries they visited, thus they were of an informal, if not illegal, character. If the sphere of informal economy in the USSR was vast, some scholars claim it even grew in the aftermath of the empire's collapse (Eder et al. 2003: 4).

Some data could be acquired from indirect sources like migration offices, tourist agencies, cargo companies and, only lastly, tax offices. The problem here, however, was that only a portion of the people going abroad were *chelnoki*. In addition, in the early

27 One language and translation problem should be noted here. Petty traders do not describe their activity as *biznes* (business) but as *torgovlya* (commerce or trade). The term *biznes* is reserved for economic activity on a much bigger scale; a *biznesmen* is someone wealthy and prosperous.

1990s citizens of successor states of the Soviet Union still carried Soviet passports. Therefore, statistics of the countries that these people had visited are inaccurate. It was only in 1995 that the Central Bank of Russia worked out an extensive methodology to measure the phenomenon of international shuttle trade, with the role of imported foodstuffs and clothes within the national turnover of these goods as a key factor. In any case, the size of the gray economy linked to petty trade was considered to be enormous. That applies to the 2000s as well (cf. Eder et al. 2003: 4–6; Kapralova and Karaseva 2005: 407–409; Yakovlev el al. 2007: 85).

One thing can be stated for sure: the phenomenon of *chelnoki* was massive. Estimates that were made vary significantly. Some scholars, for instance, consider the years 1993 and 1994 to be the peak of that activity, with 10 million people engaged in it (Eder et al. 2003: 5). Others, in turn, confirm that data, but shift the peak to the second half of the 1990s and assess the number of shuttle traders at 18–30 million people at that time (Kapralova 2005: 402; Mel'nichenko et al. 1997: online). These estimates differ depending on whether one counts only the traders themselves or includes into the number also people engaged in various services tied to this trade. For the sake of simply making the scale of this clear, if they were 30 million people working in trade, that would make up 35% of Russia's working population at that time (cf. Vasin 2008: 49).

This army of people played a crucial role in supplying Russia with consumer goods. Estimates say that in 1995–1996 around 30% of Russian import of the goods was performed by shuttle traders. For instance, traders brought altogether around 30% of audio and TV equipment, 50% of leather products, and as much as 70% of clothes and shoes imported into Russia. Scholars assume that in some branches the size of the shuttle trade exceeded official international sales by several times. In 2004 the share of individual traders in Russian import was still over 20% (cf. Kapralova and Karaseva 2005: 404; Prikhodko and Volovik 1997: 17; Yakovlev et al. 2007: 85). All in all, the numbers show that generations of socialist rule have hardly impaired people's "ability to act and react like optimizing entrepreneurs" (Hann and Hann 1992: 3).

If one thinks about the volume this trade had and its role for the Russian people, it is unbelievable to imagine that at the early stage all these goods were carried to Russia just in one's hands. That fact gave another name that is usually used for Russian shuttle trade – the suitcase trade. *Chelnoki* brought the goods in nylon bags. It was only later that professional travel and cargo companies appeared. The tourist agency of Valentina and Pyotr, for instance, planned the trips but did not cooperate with any cargo company, so it was completely up to the traders to transport the goods they bought. In the early 1990s everything was done spontaneously and on one's own.[28]

There were several causes of the outburst of petty trade. Shortages were one of them. One has to count here also the fall of Russian industry, which resulted in deep recession and widespread unemployment. Moreover, petty trade is a kind of business that is relatively easy to enter into. One does not have to possess large starting capital. In the first half of the 1990s the costs of the majority of shuttle trips – including

28 As one study reveals, in 2003 the average load that a single trader transported weighed 126 kilograms (Kapralova and Karasayeva 2005: 405). Though at that time the majority of traders sent their goods by couriers, some of them still relied only on themselves. One can suppose that in the early 1990s the size of the loads was approximately the same and everything was carried by hand.

the goods bought in Russian, and travel and accommodation costs – did not exceed several hundred dollars. At the beginning petty traders hardly paid any taxes, Gaidar intentionally did not impose any custom fees on goods brought to Russia by private individuals or entrepreneurs. *Chelnoki* did not even have to register their activity, which on one hand encouraged people to start the business, but on the other vastly expanded the gray economy around the trade. Last but not least, people were finally allowed to travel out of the country and the ruble became internationally convertible. Gaidar's policies of keeping a strong currency rate was another incentive to import goods from abroad (cf. Kapralova and Karaseva 2005: 400; Mel'nichenko et al. 1997: online; Poretskina 2006: 24; Prikhodko and Volovik 1997: 17; Yakovlev et al. 2007: 88).

❖ The economic geography of new Russia

With their cargo the suitcase traders travelled across long distances, using trains, buses, ships, and planes. There are over 5,000 kilometers from Krasnoyarsk to Poland, for instance, but some of the trips were even longer. A distinct feature of Russia lies in the simple fact that the country is so vast. The wide range of shuttle destinations is also a result of this circumstance. Obviously, a trader from the Far Russian East preferred to get goods in China, South Korea, or Japan. For people from the European part of Russia, in turn, it was easier to travel in a western or southern direction, so the traders would go to Finland, Poland, Germany, Italy, Greece, Turkey, and Syria. One has to consider here the so-called near abroad (*blizhneye zarubezhe*) as well, i.e. the former Soviet republics. States like Belarus or Ukraine became important intermediate spaces through which goods from Western Europe were delivered to Russia. However, many traders travelled to countries quite distant from the Russian border, e.g. the United Arab Emirates or India.

An important travel destination for the *chelnoki* was Poland. They were present at markets in literally every town and every city, but the most popular was the Tenth-Anniversary Stadium in Warsaw. In 1989 the stadium was turned into a huge outdoor market called Jarmark Europa (Europe Fair). With several thousand registered and unregistered traders, the market was one of the largest in Central Europe.[29] The Stadium was invariably listed among the ten biggest enterprises in Poland, though its real annual turnover was difficult to assess. Estimates for 2001, for instance, fluctuate between 125 million and 3 billion US dollars. In addition, studies say that the livelihood of up to 500,000 Polish citizens could have been supported by the market's existence, which shows how this trading was important not only for Russian shuttlers but also for the local population. Beside completely legal activities, the Stadium was a place for unregistered labor and clearly criminal activities such as the sale of pirated CDs, smuggled cigarettes, unlicensed alcohol, and weapons. To make a long story short, serious money was involved and perhaps that was the reason why, after many earlier attempts had failed, the market finally shut down only in 2007. It was only the organization of the European Football Championship in 2012 and the need to build a national stadium on this site that managed to drive the traders out of the market (cf. Cichomski 2010; Pawelczyk 2001: online).

---

29 The biggest open-air market in Europe was probably the Cherkizovsky market in Moscow with around 100,000 people employed in its heyday. The market was definitely the biggest in Russia (cf. Mörtenböck 2007: 81–82).

Nevertheless, it was perhaps Turkey that played the most important role in the new Russian economic geography. Studies estimate that in the mid-1990s from 40% up to 55% of suitcase traders went there (Eder et al. 2003: 5–6; Kapralova and Karaseva 2005: 404–405; Prikhodko and Volovik 1997: 18). The Laleli market in Istanbul was a mecca for shuttlers. A survey conducted by the Turkish Central Bank indicated that in 1996 roughly 1 million people from the CIS countries visited Turkey with the aim of trading. Around 70% of deals closed by them took place at Laleli, so the whole practice was very concentrated in terms of place. At its peak in 1995 the turnover at the market reached an estimated sum of 10 billion dollars (Eder et al. 2003).

The whole business was so profitable for the Turkish side that the Turkish government introduced policies to attract tradespeople from the CIS countries.[30] Specialized travel agencies – both Russian and Turkish – appeared and supplied traders with a range of services. They planned the travel routes for them, provided them with visas, and made agreements with particular hotels, shops, and even producers.

However, even without special support programs, open-air markets mushroomed in the regions neighboring the border with the former Soviet Union. The markets were called Russian, although in fact the majority of tradespeople there were Georgians. The attitude of the local population to the markets was of an ambiguous character: although the locals profited from the trade, the vendors were often disliked and described as *dirty*. This characterization had a moral connotation and concerned prostitution. Selling at a market in Turkey and selling sex were closely linked in the case of tradeswomen from the former USSR. There were two basic reasons for this: the tradeswomen's need to earn money and the demand among local men for such services (Hann and Hann 1992: 4–5).

Another important country for the shuttle trade and comparable to Turkey was China. Its role was steadily rising during the 1990s and the 2000s. Again, one has to take into account the geographical factor. In 2004 in Moscow around 25% of imported consumer products came from China, but in Khabarovsk Krai they constituted as much as 90%. Just like Turkey, China implemented several new regulations to ease the activity of Russian traders. They were exempted from any visa requirements or customs fees (Kapralova and Karaseva 2005: 403–409; Yakovlev et al. 2007: 95–96). Subsequently, Chinese traders started to sell their national products in Russia themselves. Moreover, Chinese goods also got to Russia through tradespeople from the Central Asian states.

❖ The 1998 crisis and the evolution of petty trade

In the course of time, Russian international shuttle trade underwent significant changes. As some scholars put it, it began to "take on more civilized forms" (Kapralova and Karaseva 2005: 410). However, it would be perhaps better to say that it started to reshape and to shrink.

The financial crisis of August 1998 played a crucial role in that process. The crash in the Asian markets coincided with excessive widening of Russia's budget deficit. On

30 There was for instance a special program designed for the Black Sea harbor of Trabzon. Any trader who came to the city by ship and bought at least $2,000 worth of goods, received reimbursement of travel and accommodation costs. The amount that had to be bought was $4,000 if a chelnok came by plane (Kapralova and Karaseva 2005: 403).

August 17 the Russian government issued a 90-day moratorium on foreign debt payments, essentially declaring the country's insolvency. This led to a jump in inflation and to devaluation of the ruble. Within a week its value in relation to the US dollar had dropped by several times (cf. Aslund 2002: 251; Smorshchkov 2013: online). It was not unusual for the ruble to suffer a sudden and large loss in value (cf. Aleksashenko 2009: 37ff.). This devaluation, however, was one of the most serious in the history of the new Russia and it brought enormous consequences with it.

People lost their savings again because of the sudden rise in inflation and because of the fact that about half of Russia's commercial banks went bankrupt (Aslund 2007a: 179). Actually, everyone who had any outstanding payments in foreign currencies went bankrupt immediately. This included the traders, who often bought goods on credit from their foreign partners. Such credit, like almost all transactions made by the *chelnoki*, was always in dollars. Even for those who did not go bankrupt, acquiring goods abroad became simply too expensive.

As the Russian economy recovered after the crisis, some traders reentered the business. Meanwhile, however, other important shifts took place. Because of the crisis and the rapid fall in turnover, cargo companies had to compete for clients. In practice, this meant lowering their carriage fees. Letting somebody transport one's goods became more profitable than transporting them on one's own. Already in 2004 the vast majority of imported goods were bought in Russia (Yakovlev et al. 2007: 94). These goods were brought by wholesalers or simply ordered from abroad and delivered to Russia.

In other words, the former *chelnoki* began to become domestic traders. Shuttle trade remained international in the borderlands, but the majority of traders did not personally go abroad to get their goods, rather they acquired them in Russia. The people who returned to business after the 1998 crisis or even started their activity at that time mostly came from less developed regions. They travelled to big urban centers where wholesalers operated (especially Moscow), bought the goods there, and then sold them in the provinces. There the competition was less intense and the costs of business activity were lower. Thus, the possibilities to stabilize one's enterprise and to make a profit were better. This situation has not changed up to the present day.

Nowadays none of the traders I talked to in Krasnoyarsk personally traveled abroad to get the goods and transport them in nylon bags on his or her own. People might travel to establish business contacts, like Pyotr, who goes to China from time to time. Kyrgyz traders, for instance, sometimes go to Kyrgyzstan to check what new products textile factories in Bishkek have introduced into the market and to make orders. The opportunity to reunite with their families, however, is probably much more important to them than commissioning the next batch of goods. After all, they can easily place their orders from Krasnoyarsk. The majority of tradespeople in Krasnoyarsk get their goods locally from wholesalers. Some order from Moscow. Interestingly enough, the latter often order Chinese products, i.e. products that presumably earlier passed through Krasnoyarsk. If traders travel to buy products, they go to Novosibirsk where some large wholesalers operate for the whole of Siberia. On the other hand, many people visit Krasnoyarsk itself to get goods. They come from different districts of the Krai, some of them quite distant like Norilsk, but from other Russian regions as well such as Khakassiya, Tuva, and Irktutsk Oblast.

There were several reasons why the *chelnoki* gave up their trips. One reason were the problems that were inevitably tied to long journeys and crossing of borders. Obviously, it is more comfortable to let a cargo firm transport your goods for you, but it is much easier not to travel at all. Little by little, after some traders had established contacts with their partners abroad, this started to be the case. An important aspect in this respect was the improvement of transport infrastructure and the appearance of large wholesale companies. Going on one's own stopped being economically viable.

One should also consider the question of competition. It was the time after the 1998 crisis when foreign supermarket and hypermarket chains entered Russia. These initially started to operate in Moscow and other European regions of Russia, but they soon began to spread eastwards. Over the course of time some Russian retail networks appeared as well. The problem of organizing a supermarket is that it is a sophisticated task in terms of logistics and management. The necessary knowhow had to be brought to Russia from the West. Running a supermarket demands an advanced system of deliveries which, in turn, requires a developed transport infrastructure, above all good roads. That is why to this day the biggest retail networks in Russia practically do not go any further east than Novosibirsk.[31] Smaller networks, however, were rapidly developing also in the eastern regions of the country. Over the years the chain stores started to pose a serious threat to petty traders and now are among their biggest problems (cf. Goskomstat 2013: online; Radaev 2003; Radaev 2005, Radaev 2011a).

What is more, in the course of the 1990s Russian leaders started trying to seize control over the *chelnoki* and petty trade in general. Obviously, no state takes kindly to not knowing what its citizens do for a living or what and in what quantities they bring into the country. The fact that suitcase traders did not have to register and did not pay any customs fees meant for the state budget perhaps not losses, but at least revenue that could be earned. Individual imports turned into an important branch of the economy, but it did not bring to the budget anything at all. The first limitations on international shuttle trade were implemented in 1994. Further restrictions were enacted in the following years.[32] Interestingly enough, these legal restrictions did not discourage the *chelnoki* from their activity, nor did initially the income tax that was subsequently imposed, although it became increasingly uncomfortable for traders as time went by.

The *chelnoki*'s activities were profitable: in the early 1990s a single trip paid for itself and gave a trader around 100% of profit from the goods she bought (Kapralova and Karaseva 2005: 402). A 100% profit margin in trade is a high rate. Relatively high earnings were not the only factor which prevented the customs fees from harming the shuttlers so much. After all, it was a particular customs officer who assessed the value of an imported load, and with a particular customs officer – as Russians say in such cases – *vsegda mozhno dogovoritsya* (you can always reach an agreement). However, in order to do it, you need to have good connections, because the agreement has to be safe for both sides.

31 As for 2013 this was, for instance, the case of Auchan and IKEA stores, despite the rapid development of the two corporations in Russia.

32 Since 1994 an entrepreneur or a private individual could bring into the country duty-free goods worth up to 5,000 dollars. Further restrictions were introduced in 1995 and 1996. The value of goods brought duty-free was lowered to 1,000 dollars and to 50 kilograms of weight. For loads of up to 200 kilograms and worth less than 10,000 dollars a fee of 30% of the goods' value was collected.

Until the 1998 crisis the value of goods brought during individual shuttle trips constantly grew (Kapralova and Karaseva 2005: 402). The crisis, however, stands as a dividing line. Since that time one can talk about a decline of petty trade in Russia.

The women who saved Russia

Andrey Nechayev explained the reasons for liberating trade at the end of 1991: "We considered it crucial for the people to bring consumer goods into Russia. And this they did all right and, yes, aesthetically it was off-putting. Do you remember the *chelnoki*, those miserable women, each wearing five overcoats? Only a Russian woman is even capable of such things: five huge bags in each hand! But they were the ones who saved the market, and for this they deserve acknowledgement and all the gratitude" (Nechayev 2011: online). Indeed, Russia's savior was a shuttle trader, a female shuttle trader to be precise.

Surveys confirm Nechayev's statement: it was mainly women who engaged in suitcase trade. For instance, they constituted approximately 2/3 of the *chelnoki* that went to Turkey. This stood in contrast with Turkish shopkeepers in Laleli who were all male. However, they often hired Russian women in order to have better contact with their customers. Most of the female shuttlers had higher education, which has remained a somewhat characteristic feature of Russian petty trade to this day (Eder et al. 2003: 10–12).

The tradeswomen needed to have a lot of energy and self-determination, as they had to overcome obstacles such as travelling long distances, coping with protection rackets, carrying heavy loads, and passing state borders with those loads. In addition, a person who travelled in the middle of winter from Krasnoyarsk to Istanbul, for instance, was exposed to a temperature difference of up to 50°C. Climate changes are also something one has to get accustomed to. "No, I won't go myself for pullovers to Turkey. I'm old now," said a former *chelnok* in an ethnographic interview (Eder et al. 2003: 12). She was 44 at that time, which serves as indicator of how demanding the shuttle trips must have been.

One of the female shuttlers was Anya, a friend of Konstantin (B2) and Yuliya (B3) described in Chapter 3. She had a good position in the biggest hotel in Krasnoyarsk. Her husband danced in a famous Siberian dancing ensemble, which in socialist times was a much desired workplace, as the group often travelled abroad. After the fall of the USSR they both lost their jobs and started trading in shoes. To get the goods, they travelled to Poland. The trips were challenging. The work in Krasnoyarsk was exhausting as well, since they traded outdoors, sometimes in freezing cold. The efforts sapped the health of Anya's husband: after a long working day of carrying heavy packages, he suffered a heart attack and died *right in the street*. Anya was left with two daughters to support. She continued her work, but as over the course of time travelling to Poland became more difficult and less profitable, she began to go to China. She went on the trips together with a few of her female friends and the group cooperated in buying goods, transporting them, and getting across borders. All that cost her a great deal of effort, but she managed to survive the hardest time and to bring up her daughters. At a certain point she had to quit working because of serious health problems, however.

❖ *The market feeds everybody*: the story of Lyuba

Before she started to trade at the Slavyansky market in Krasnoyarsk, Lyuba (T59) used to work in a textile factory. She held a degree in economics and worked as quality controller there. However, the factory was shut down and Lyuba had to look for a new job. She came to Slavyansky in 1996, choosing the market because she lived in the neighborhood. As a matter of fact, she had sold things there before. She would go to the railway station and wait for the Beijing-Moscow train, which at that time was packed with Chinese clothes for sale. Lyuba bought some and then stood until she sold it, either at the Slavyansky or in a farm town 80 kilometers outside Krasnoyarsk, where she was born and where her mother still lives. She worked as a salesperson for various people, first selling clothes, then sweets, then cigarettes, then tea and coffee. The disadvantage was that her employers did not want to pay her pension fund contribution, so in 2001 Lyuba decided to start on her own and became an entrepreneur. She still sold cigarettes for an employer, but just across from that she had her own stall with clothes. In 2007 she changed her focus again, for more and more Kyrgyz people were coming to the Slavyansky and all of them traded in clothes, so the competition grew stronger. Therefore, Lyuba switched to foodstuffs and started to sell groats, macaroni, instant soups, and so on.

Lyuba's husband Gennady worked in a harvester factory, once one of the biggest enterprises in Krasnoyarsk, which still operates today, albeit in a very limited scope. In the beginning he did not want Lyuba to work at the market. He tried to persuade her to drop it, because he perceived commerce as something shameful. In the course of time, however, he started to accept it and supported Lyuba in her work, because he was paid only every few months, while Lyuba's job brought regular money, which was crucial in sustaining themselves and their son.

"*Trade let us survive*," Lyuba emphasized several times. "*The market feeds everybody*," she said, while pouring groats for *her children*, i.e. the pigeons at the market, the pigeons that other traders at the Slavyansky fought against. However, nobody would quarrel with Lyuba about that, because she was well liked and respected.[33] She also gave tinned food to Vasya, the market's dog. That was her position: the market feeds pigeons, dogs, and traders themselves. Since she survived thanks to it, she was obliged to let others survive too. *To survive*, Lyuba explained, is to get by possessing very little means that scarcely fulfill just one's basic needs. In the 1990s, for instance, her family could only afford to pay electricity bills from time to time, so the light was sometimes cut off. On the one hand then, electricity is something you can do without. On the other hand, without electricity you are barely surviving, because it makes your existence tougher.

---

33 I witnessed once that pigeons Lyuba fed defecated on one of her fellow trader's hair. The women treated it with humor, however. After that Lyuba promised not to feed the pigeons any more, but she managed to keep the promise for just a few days. She started to feed birds just behind the market gates, which in fact made little difference. The pigeons were left for a while, but then they came back again to "*their mother*". Other traders seemed to accept it: Lyuba was simply just like that. They were ready to tolerate the pigeons in order to maintain a good relationship with her. Even a Tajik trader, who sold dried fruits and nuts, and for whom pigeons were a real curse, gave Lyuba sunflower seeds for the birds for free.

Then Gennady lost his job. He tried to find another, to work for instance as a taxi driver, but "*it was not his thing*". He did not want to work at the market, either. In the end he "*gave up*" ("*on sel*"). Later he was diagnosed with widespread lung cancer and died within six months. As Lyuba said, Gennady lost his will to live. She added also:

> "*We think men experienced the perestroika time much harder than women. I observed it with my husband's colleagues and with the neighborhood men. They often started to drink. A man is stronger physically and a woman is stronger spiritually. That is why they should be together.*" (T59)

One can conclude that for Gennady it was difficult to face the situation of unemployment. In his opinion, a man should work at a factory, should do physical labor by producing something. That is why he rejected working at the market. She, in turn, managed to find a job in trading and in the course of time to even open her own business. Thus, it seems that in this mode of thinking the difference between men and women lies in the attitude towards labor activity. There are different expectations in terms of one's career and success. While the only respectable opportunity for a man is a professional career in industry and to be the main breadwinner, a woman can find fulfillment in many spheres, even a non-professional one (Lyuba took to trading at the market despite holding a university degree). "Men's status in the household is defined by their role as primary breadwinners and that when this role is lost, it is very difficult for them to define a new position in the domestic sphere," as Sarah Ashwin and Tatyana Lytkina wrote about male Russians (2004: 190–191).

Generally speaking, in the reality of the postsocialist systemic transformation, labor often seemed to be the key factor for male self-identification. This was especially the case whenever an ethos of physical labor defined whole social classes, like peasants (cf. Buchowski 2004: 189). That is why it was much more difficult for a man to cope with unemployment. In other words, women displayed better adaptation skills. In a situation of rapid social change these skills enabled them to better withstand various crises, while inversely those crises had a more harmful impact on the psychological and physical health of men, especially in Russia.[34]

Entrepreneurship, informality and criminality in Russia

This section concerns not only petty traders but is of a more general character. I examine the strong connection of trade and entrepreneurship in Russia with informality and criminality. This connection was, on the whole, inherited from the Soviet times.

34 It is enough to mention that in 1989–1999 the life expectancy of men fell from 64 to 59.8 years of age (in 1994 it was just 58). In the same period the life expectancy of women decreased to a much smaller degree: from 74 to 72.2 years. One has also to consider that the retirement age for men in Russia is 60, so in the 1990s an average man did not live long enough to even start receiving his pension.
In addition, in 1990–1995 the number of deaths connected with alcohol grew threefold, but a man was 4 times more likely to die due to alcohol-related problems than a woman. Also, the number of suicides increased in this period by over 60% among men, but among women it went up only slightly (Ashwin and Lytkina 2004: 204–205). Estimates say that there were as many as 3.4 million premature deaths in Russia in 1990–1998 (Rosefielde 2001: 1159).

My point is that informality, i.e. operating outside the control of the state, has advantages as well as disadvantages for an entrepreneur. It is positive because one can fully use one's creativity and at the same time it is negative because one is exposed to a number of risks and uncertainties, such as organized crime. By no means, however, is the control of the state desirable for an entrepreneur. In an ideal world the state would provide its citizens with luxury conditions for the development of their own economic initiatives, though in practice Russian entrepreneurs try to avoid state structures as much as possible.

❖ *Vory v zakone* and protection rackets

One could call the 1990s until the 1998 crisis the golden age for petty traders. It was then that Yevgeny (B5), an experienced and successful businessman, made his money. During the late socialist period he was a *fartsovschik*. In the early 1990s he felt the winds of change and started to import foodstuffs and alcoholic drinks into Russia. He was a *chelnok* at the beginning, but soon he started to operate on a bigger scale. The daily turnover of his shop could be so high that counting it took hours. There were so many banknotes that he and his partners even developed an allergy on their hands. Nonetheless, even a *chelnok*, as Yevgeny put it, could earn enough to buy a really good car in a single month. Of course, the amount of money Yevgeny and his partners counted had to do with hyperinflation and a really good car in 1990s Krasnoyarsk meant a vehicle already a few years old imported from Japan. Still, the fact of the matter was that the *chelnoki* could indeed make high earnings in comparison to average salaries at that time.

Yevgeny described the general conditions of the Russian market in the early 1990s. Shortages were grave but competition, on the contrary, was weak. He mentioned also that the state did not burden traders very much with inspections, taxes, and other duties. One could say that, formally, trading was quite easy. At the same time, however, Yevgeny recounted some informal obstacles a trader had to cope with:

> *Y: You see, there used to be criminals all around. Just like that, everyone around would pay money to be able to trade. And if...*
> *K: The money went to the racketeers?*
> *Y: There was a racket all right. That's what we call criminals in Russian, "reket". Without this you would not be able to trade. The entire city was divided between groupings, and each grouping would collect the "kommerses", that's how they called people engaged in commerce in their slang, the criminal slang. And so they would collect money from everyone. Each shop, each stall, each factory, everyone paid. Everyone who... In the 90s practically 90 percent of enterprises paid to be able to work, because the authorities were unable to secure the entrepreneurs' rights, to defend them. Only criminals were able to defend the entrepreneurs.* (B5)

The *reket* has a long tradition in Russia, although the notion itself became popular in the 1990s, as a loan translation from the English word *racket* (Komlev 2006: online). In Russian it denotes mainly extorting money for protection, although it is also strongly connected with other branches of organized crime, such as robbery, theft, kidnapping for ransom, deals on the black market, murder, and so on. The *reket* is closely tied to the term *krysha* (roof), meaning a person or group that secures protection. There are many kinds of roofs; besides the criminal *krysha*, there is also one provided by the police, a political *krysha*, an administrative one, and so on, but in everyday usage they are all described with one collective term. It is the system of patronage which matters here, and whether the patron is a legitimate or a criminal power structure is of lesser importance (cf. Humphrey 2002b: 78–79). What all these sorts of *krysha* have in common is that one has to pay for them, using money or services.

Secret organizations and gangs of professional criminals existed in Tsarist Russia and in the Soviet Union as well. In the Soviet Gulag system a whole prison culture stemmed out of the tradition and it is there that most criminal associations formed. The most famous of them were so called *vory v zakone* (thieves in law), which can be translated as criminals who obey the thieves' code. The criminal law was not only parallel to state law but completely opposite to it, since one of the main principles for a *vor* or even a candidate willing to reach this elite rank was to not collaborate with authorities in any way, shape, or form.[35] In the course of time, the code and the criminal hierarchy that was connected to it spread across the entire Soviet Union, so that even criminals who had never been in prison followed it (Humphrey 2002b; Varese 2001).

Organized crime started to flourish during perestroika. In the 1990s, as Yevgeny pointed out, racketeering was a general phenomenon. Bandits who extorted protection appeared in virtually every story of entrepreneurs from those times. The simple truth was that in most cases if you wanted to run your own business or sell at an outdoor market, you had to pay for protection. In Moscow even old women who sold sunflower seeds at the pavement had to pay for their own *krysha*. Entrepreneurs bought protection that was really protection, one could say, from other bandits. In other words, the system of racketeering was about working under constant threat: bandits protected you from other criminals, but if you refused to pay them, you risked experiencing their own violence. Nonetheless, since perhaps every business in 1990s Russia was a highly volatile occupation and nothing could be certain, there were entrepreneurs who had to pay two or even three different criminal groups at the same time.

❖ Property and demand for its protection

Widespread demand for protection – as Federico Varese argues – appeared in Russia as a result of the rapid and large-scale privatization of property.[36] According to him, privatization itself was not enough to produce a well-functioning market economy, because the cornerstone of such system is not private ownership itself but the presence of property rights. In the Hobbesian state of nature, an individual's property is insecure because an individual is unable to protect her life and assets from the hostile attempts of other people. Therefore, people decide to conclude a social contract and cede part of their rights to a sovereign authority, i.e. the state. An important

35 For instance, the code strictly forbade serving in the army. During World War II this rule lead to a serious break among criminals in the Soviet labor camps. The prisoners were offered a chance to join the Red Army and to wipe out their records by fighting the enemy. Many of those who accepted the offer found their way to camps again after the war ended. This resulted in so-called *bitch wars* (*suchye voyny*), wars between thieves-in-law and those who *went bitch* (*ssuchilsya*), i.e. broke the thieves' code. It is estimated that thousands of people were beaten or killed in the 1950s' prisons for that reason (Humphrey 2002b: 106–107).

36 By the beginning of 1996, as many as 72% of Russian large and middle-sized enterprises and 82% of small shops and retail stores were privatized (Varese 2001: 213).

consequence of ceding one's rights to a state, however, is the problem of protecting individual property from the state itself. When the state is non-impartial and predatory towards its citizens – and this, by the way, is also the case of the Hobbesian ruler who is above the law – two tendencies can be observed: the demand for state protection services falls while attempts on the part of citizens to influence the state's rule increase (Varese 2001: 23–25).

In 1990s Russia the rules of governing were far from being predictable, property rights were not guaranteed at all, and the state was highly non-impartial in its policy towards various social classes. In a system described as *bandit*, *oligarch*, or *nomenklatura capitalism* it was the state official who became the main entrepreneur and politician. The most valuable prerogative the state official had was the right to take administrative decisions (Klyamkin and Timofeyev 2000). Citizens wishing to influence the state ruling, i.e. to obtain decisions advantageous for them, had to corrupt the system. In other words, in order to run their businesses Russian entrepreneurs often faced the necessity of bribing officials. As the informal economy is of an ambiguous status, this could be a burden – when it came to receiving necessary licenses, for instance – but it could very profitable as well. In the course of the Russian transition the most lucrative opportunity turned out to be influencing administrative decisions concerning privatization.

Over the first half of the 1990s state property was being privatized on a massive scale. The process reached its peak in 1995 with the loans-for-shares program (*zalogovye auktsiony*), endorsed by prime-minister Anatoly Chubais. As a result several of the largest national industrial assets were sold at very undervalued prices, including the oil companies YUKOS, Sibneft, and LUKoil, and such giants as Norilsk Nickel. Yury Afanasyev supposes that the loans-for-shares privatization might have been the biggest fraud in world history and sees in it the source of prodigious oligarch fortunes. According to him, it was because of this program that a new oligarch class appeared and gained a great influence on national policy (Afanasyev 2001: 166–168). Anders Aslund, in turn, claims the loans-for-shares privatization was an unmitigated success. Indeed, the companies were bought at only a small part of their real value, but the new owners made them – in his view – extremely profitable, which entailed bringing large revenues into the state budget (Aslund 2007a: 267–268). Irrespective of the assessment of the results, loans-for-shares privatizations were mostly fictional in the sense that the auctions of industry assets were rigged. This caused the whole process to be one great fraud and that was in some cases admitted even by its main actors (cf. Overchenko 2011: online).

Another fact is that some oligarch fortunes were later renationalized. The stories of Mikhail Khodorkovsky, Boris Berezovsky, and Vladimir Gusinsky proved that even the privileged members of the oligarch class could not be certain of their property. The market that was introduced in Russia was not free from the very beginning, but in the course of time the state's grip on it only strengthened. The state, however, was not public as it was during the Soviet times, rather it was privatized by a new *nomenklatura* that consisted of state officials and major businessmen strongly linked to each other. Thus, postsocialist property clearly can be called a fuzzy one (Afanasyev 2001: 210; Verdery 1997).

❖ The substituting of the state: the story of Alik

Alik (T12), an Azerbaijani tradesman, was 42 years old when I met him at the Slavyansky market in Krasnoyarsk. He traded in fruits and vegetables together with his friend Fedya (T9), another Azerbaijani. They had made an agreement: Fedya had a stall at the market, but he was out of money. Thus, it was Alik who invested in buying the goods. They worked together and their idea was that they would share the profits equally.

However, the plan did not work out. A few years earlier the Slavyansky had been a busy place, where good money could be made because customers came in droves. In 2012 the situation looked quite different. Because two supermarkets had opened in close vicinity to the market, and because of the ensuing financial crisis, there were much fewer customers coming to the market. As a result, many entrepreneurs were forced to give up their businesses.

As it had quickly turned out, for Alik and Fedya there were practically no profits to share. Besides the competition posed by the supermarkets, there were many other tradespeople at the market who sold fruits and vegetables. As a consequence, the two men had to close their stall. I did not see Alik for a long time afterwards and I thought this was the end of the story. But then I ran into him at another urban bazaar. He told me that his engagement with the Slavyansky had been a mistake. It was an unlucky period in his life, he was confused and he did not know how to earn a living and provide for his family. That is why he entered into cooperation with his friend. Their plan misfired, but Alik was quite happy about it, for the failure had made him start a new business.

Alik moved with his family to Krasnoyarsk in 2010. He had been a successful businessman in Azerbaijan before. He imported wood from Russia and used to come to Krasnoyarsk from time to time. His business went well until one day, a whole previously paid-for shipment of wood did not arrive. These several wagons were worth, he claimed, 240,000 US dollars. As it turned out, Alik's business partner – a Russian man from Krasnoyarsk, who until that time had been a reliable person – failed him. Alik came to Krasnoyarsk and found the man. Initially, he tried to resolve the problem peacefully. Later, since the man just tried to trick him, Alik gathered a few of his compatriots and kidnapped him. They kept him in a basement for a few days, trying to extort the money. They soon found out, however, that the former partner lost almost everything partying and gambling. They eventually let the man go and the deal was that he would repay Alik's money in installments.

The new business Alik started was importing male underwear from China. As he told me, his distant cousin had a textile factory there. At the beginning Alik ordered a relatively small batch, just for a try. He proved successful, selling the merchandise within a few days. He then started to order larger shipments and his business grew. What was important to Alik was that he worked in wholesale trade. As he stressed, he no longer needed to spend whole days in a booth and to quarrel with customers about every ruble.

What is striking in Alik's way of solving the problem is his relying only on himself and his personal contacts. His conduct can be presented as an expression of trust in his fellow countrymen on the one hand, and mistrust towards state institutions on the other. One has to keep in mind the very specific conditions of running a business in Russia. As Galina Lindquist observed, "The ethos of contract, the cornerstone of the business in the West, is by no means unknown to Russian entrepreneurs" (2006: 205).

In general, in the Western business culture there is a strong moral obligation to fulfill the contract and to be loyal to business partners. The trust that the other party will not deceive you is strongly supported by legal sanctions. Such sanctions did not work in 1990s Russia. Even if appropriate legal regulations existed, in practice their application was selective. After all, "the members of law-enforcement agencies in situ are human beings employed by the State, and to survive, they have to supplement their meager salaries with private contributions from interested parties" (Lindquist 2006: 205). Alik did not secure the implementation of the law by bribing a state official. He decided to use violence and thus turned from an entrepreneur and business partner into a racketeer.

Under the Russian conditions of running a business, hiring private protection was in many cases the best way to guarantee that the other party would carry out the contract. Racketeers then not only protected one's business but also served as means of dealing with one's partners or customers. In 1997–1998 Vadim Radayev conducted a wide-ranging survey among managers of all kinds of Russian enterprises: small, medium-sized, and large ones, both private and state-owned. Among small private enterprises Radayev singled out those working in – as he called it – risky market sectors: wholesale and retail trade, public catering, consumer services, and finance. Basically all of the interviewed managers who worked in those sectors considered breaching contract a frequent phenomenon and 70% of them constantly experienced such practices. In the case of a business partner breaching the contract, 57% of them would try informal negotiations, 20% would use violence, and just 13% would bring the complaint to a court of arbitration (Radaev 1998: 149–151; cf. Varese 2001: 51–53).

What draws one's attention in Radayev's survey mentioned above is not only petty entrepreneurs' lack of trust of in their business partners but also their mistrust towards state structures. This group was least likely, among entrepreneurs in general, to report extortions and threats to police and bring cases to court, on the one hand, and most likely to hire private security companies or criminal groups on the other (Radaev 1998: 150). In other words, Alik's conduct was characteristic not only of migrants. Russian small entrepreneurs in general did not believe that the police and prosecutors could help them at all. This seems to be a result of the highly ambiguous status of Soviet and Russian law-enforcement bodies; in many cases it was the *militsionery* (police officers) who were to be feared more than the criminals.[37]

Tanya, a women who worked at the Slavyansky market in Krasnoyarsk for several years beginning from the mid-1990s, told me that she had never been persecuted by the usual bandits. The only racketeers who extorted money from her were police officers. They were constantly coming and checking every possible document, license, and so on. They could take a document from a trader one day and the next morning punish him or her for not having it. If a criminal harmed a trader, the trader could try to seek help from the police. But if the perpetrator was a police officer, the trader could not complain to anybody. It is also the serious quantity of wrongdoings committed by members of law-enforcement bodies that causes so many crimes in Russia to go unregistered. People do not report crimes to a police force that is perceived as corrupted, passive, and ineffective (cf. Aslund 1997: 81–82; Overchenko 2011: online). Because

37 According to the reform of MVD structures introduced in 2011 the *militsiya* was substituted by the *politsiya*.

of all the abuses committed by *militsionery*, to this day they are deeply despised in Russian society. In colloquial language they are called *menty*[38] or *musory* (rubbish).

The level of criminality in the late 1980s and 1990s rose significantly all over the postsocialist world. In Russia the rise was dramatic (cf. Aslund 1997: 80–82; Aslund 2007b: 242–244). In 1990s Russia state structures were weak but the criminal ones, on the contrary, were robust and in answering the demand for protection they often substituted for state institutions. Although the functioning of state institutions had significantly improved after the 1990s, Alik behaved as if those times never ended. He did not even consider reporting the case to the police or the prosecutor's office. He was convinced that those institutions would not help a migrant, even though he is a Russian citizen. That is why he took action through informal channels and why his actions were so vicious.

❖ *Bespredel*

The earnings of a petty trader in the 1990s could be relatively high – that is one side of the story. The other is that tradespeople usually remember those times with dread. A tradeswoman who had been working in trade for over 20 years told me:

> *They would kill people then, all those criminal groupings. You were literally afraid you would lose your life. The very moment you started anything, any kind of business, they would start watching you. And it was not the state that controlled it. It was simply terror, terror and violence [...] The times when trade had just started, when it was in its earliest stage – those times were terrifying. If you refused to pay your tax, they would come and warn you the first time, they would come and warn you a second time. The third time they would set fire to your shop and the goods. Or it could get even worse – there were murders, you know [...] This started in the early 1990s, to be precise, and after 2000 it somehow got more peaceful. There were murders, but only at top levels. But from the 1990s to 2000 – this was a horrible time. We were constantly afraid.* (T5)

Another tradeswoman said briefly:

> *These were the dreadful 90s, no doubt about it, this "reket", this "bespredel".* (T49)

A notion, which is often used to describe the 1990s' Russia, is *bespredel*, the noun and the neologism coming from the adjective *bespredel'ny* (limitless, boundless). Lindquist recalls that one of the metaphors for living under the Soviet regime was *prison*. It expressed the feeling of being subjugated and constrained. In the 1990s a new term came into use: *jungle*. It was even scarier than the previous one. A prisoner has at least a roof and walls that protect him from the cold and rain, and in addition he knows the rules of the place: "Many people had resented and ridiculed the old home, but they knew how to live in it; suddenly, instead of that ugly, decrepit, but so livable home there was a *jungle*, and many people started to doubt their very ability to survive in it" (Lindquist 2006: xiii). *Bespredel'ny* connoted the limitlessness, lawlessness, and

---

38 The etymology of the word *ment* is not entirely clear. Some authorities derive it from the Polish word *menda* (police officer, guard, informer), others from Latin *mentula* (penis), or see in it abbreviation of *mentally deficient* (see commentaries and discussions at: http://ru.wiktionary.org/wiki/мент and http://www.slovoborg.su/definition/мент). Regardless of the etymology, the connotation of the term is clearly negative.

cruelty of the new reality that appeared after the fall of socialism. Along with the notion of *bardak* (brothel, disorder) – used in similar contexts – it conveyed the perceived savagery of human nature which was unleashed in the chaos of the new times (Lindquist 2006: xiv; Nazpary 2002: 2). These new times have been also compared to the historical period of *smuta* (the Time of Troubles).[39]

❖ Coping with racketeers: *it's only in our Russia that they can sell air to you*

There were different ways of coping with racketeers and it seems that a lot depended on one's personal capacity and skills. Konstantin (B2) – the *fartsovschik* and the *shabashnik* I wrote about in Chapter 3 – continued to work in construction in the 1990s. At the same time, he started to supply companies working in the northern parts of Krasnoyarsk Krai with special equipment, and he does this still today. Racketeers frequently demanded that he pay them. The situation became unbearable because the bandits used to come very often. One time when they came, Konstantin asked them to wait. He went to another room and called his friend from high school. The mate was a judo master and this circumstance enabled him to establish his own gang. He came with his fellows to Konstantin's office and beat up the other bandits. Since that time Konstantin was not pestered anymore, which by no means meant he did not pay for protection. His high school friend, despite their friendship, demanded a payment. As a matter of fact, Konstantin was appreciative of this, as the amount of money extorted was smaller and the relationships with the new protectors were much nicer.

Konstantin as well as Yevgeny ran prosperous companies. Their experience, gained in the informal sector during socialism helped them not only to launch successful businesses in the new Russia but also to deal with bandits and law enforcement bodies alike. Bribing was nothing new for them, for instance. They also grew up in Krasnoyarsk and therefore had a lot of ties they could make use of in case of need. Petty traders, on the other hand, were much more prone to the *reket*. They could not afford hiring private bodyguards, and therefore they were usually in the hands of fate, having to pay gangsters for protection, or even gangsters and police at the same time. Until the present day those who trade informally in the streets either gather money to collectively bribe policemen or – if they do not manage to bribe them – work under the constant risk that officers will come and fine them. Perhaps the most vulnerable people were – and still are – labor migrants, who were foreigners and thus did not have any connections.

As Yevgeny maintained, even though in the 1990s almost every shop or private enterprise in Krasnoyarsk had to pay protection money, in course of time the situation greatly improved. Today, as he claimed, no entrepreneur gives payments to mafia structures. Indeed, vendors from the open-air markets confirmed that they are not harassed by bandits in leather jackets anymore. At the same time, however, I was told that criminals had reached another level. They extort money not from the traders one by one but from markets' administrators. These administrators, in turn, must regain this money from traders in the form of higher rents.

---

39 "A situation emerged which would hitherto be called a *smuta* but now got a new name: *bespredel*," as Zinovyev wrote in one of his books about 1990s Russia (2005: 242).

A large, 20-square-meter trading place dubbed a "shipping container" at the Krastets[40] – the huge clothing market on the outskirts of Krasnoyarsk – cost in 2012 as much as 1,200,000 rubles. The monthly rent for it was 60,000 rubles, a high sum if one concerns that the average salaries in Krasnoyarsk were between 20,000 and 25,000 rubles at that time. Nevertheless, it was not only the rents that traders were supposed to pay to the administration of the market. Traders had also to pay a so-called *guarantee fee.* As Yekaterina (T6) explained, the fee guaranteed that *your hitherto purchased place will not be sold to another person.* The guarantee fee was also 60,000 rubles and traders had to pay it every six months – in addition to the rent. It was a security company hired by the market's administration that extorted the fee.

Thus, one can conclude that vendors at the Krastets could have no certainty about their property rights. You could buy a container and it was yours, but if you did not pay the guarantee fee to the security guards – i.e. the regular racketeers in this case – you would lose the place and all the money you invested in it. Nadya, another tradeswomen from the Krastets, stated*: It's only in our Russia that they can sell air to you* (T5).

What is more, tradespeople could not be certain about the amount of fees and additional expenses. Besides collecting the guarantee fee, security guards could fine the vendors at the market for minor infractions, such as displaying goods outside one's trading space or not using the plastic bags distributed by the administration at a cost. As Yekaterina said, the guards did this eagerly because the fines went into their own pockets. It was especially migrant vendors that were vulnerable to such fines, as they were easier to intimidate than the Russians.

One can conclude that the majority of traders during their careers met with violence or were victims of extortion, whether in Krasnoyarsk or when they travelled to obtain wares. For instance, Irina (T49) went to Poland several times. She used to take a plane to Moscow and from there continue her journey by train. In Belarus there were always bandits getting on the train and warmly greeting the passengers with *The Belarusian "reket" welcomes you!* They robbed the tradespeople of a portion of their money – money that these people usually had had to borrow in order to undertake the journey. She witnessed acts of brutal violence when people did not want or did not have enough money to pay the bandits. Her position was slightly better when she traded in Krasnoyarsk. Then she tried to establish more a familiar relationship with racketeers and to negotiate with them on her payments:

> *I tried to strike up a conversation, I would say: "Let us get rich first, then we'll pay"* (T49)

This strategy was sometimes successful: payments to bandits could be lowered or postponed.

This is an important aspect. There is no doubt that the bandits were cruel and ruthless. As the first of the women quoted above said, they could burn down one's

40 *The Krastets* was a popular name for the huge clothing market that consisted in fact of two shopping malls, the *Sodruzhestvo* (*Unity*) and the *Vostochny* (*Eastern*) markets, and a few adjacent trading areas. The name came from the neighboring Krasnoyarsk Heat and Power Plant (*Krasnoyarskaya Teploelektrotsentral'*, or KrasTETs).
During its history, the Krastets underwent significant changes. Its territory was reshaped and the market itself was reconstructed several times but basically it consisted of literally thousands of shipping containers and in this respect was similar to many market in the post-Soviet space (cf. Humphrey and Skvirskaja 2009).

shop, they used terror, and they killed entrepreneurs. However, as some research demonstrates, people's perception of them was not purely negative. Racketeers can be "human", "it is possible to deal with them in a humane way", but *menty* are "real beasts" (*nastoyashchiye zveri*) according to a tradeswoman interviewed by Lindquist (2000: 343–344, 354). Apparently, although in practice they might look alike, there was a big difference between police officers and bandits. After all, it is a part of a bandit's identity to extort money, to beat, or even to kill. In a sense you know what you can expect from them. This is not the case with policemen, whose role it is to ensure the public order and to protect people from bandits, not to be bandits themselves.

Thus, somewhat ironically, bandits are somehow honest in what they do.

❖ To be a honest bandit: the story of Tolya Bykov

I was quite surprised when I first heard about Tolya Bykov, a well-known businessman and politician who settled down in Krasnoyarsk. A friend of mine told me, *he is a bandit, but a good one.*

In the 1980s, Bykov was simply a gym teacher in a small town in Krasnoyarsk Krai. During perestroika, however, he started to make a name for himself. He masterfully combined political, business, and allegedly also criminal activities and in 1998 this allowed him to become the chairman of the board of directors at RUSAL Krasnoyarsk, the biggest aluminum plant in Russia, and the second biggest in the world. That is why he was called the "King of Aluminum". He was arrested a few times. Finally, in 2002 he received a suspended sentence of six and a half years in prison for ordering the murder of a former associate and most likely fellow gangster. The sentence in itself is interesting here, though it was not even executed. Bykov brought his case to the European Court of Human Rights in Strasbourg and eventually, in 2009, he won. For committing mistakes during the prosecution the Russian state was sentenced to pay Bykov 1,000 euro in moral compensation and to cover 25,000 euro of court costs. Now he is a member of the Kransoyarsk Krai Duma.[41]

Inhabitants of Krasnoyarsk usually do not condemn Bykov for his criminal past. On the contrary, he is respected for his charitable activity. In the 1990s, when the state had no money, he generously supported hospitals, orphanages, and schools. He is also perceived as a local patriot, who tried to defend the Siberian people from corrupt officials in Moscow. Krasnoyarsk's citizens do not call him by his full name Anatoly Petrovich; for them he is rather Tolya, somebody close and familiar. Presumably, it is not only his wealth and connections that allowed him to be elected to the regional parliament. Surveys say that 38% of Russians would vote for a criminal. In the common perception, every politician is guilty of something, everyone is corrupt. Thus, the choice is not between a criminal and a just man, but between a better criminal and a worse one (Klyamkin and Timofeyev 2000: 224–225). It was because of Bykov's support for the Patriots of Russia that the party won the local government elections in Krasnoyarsk in September 2013. Krasnoyarsk was one of the few regions in the whole of Russia where the United Russia did not win.

---

41 There are dozens of press reports about Anatoly Bykov, which tell his story in detail. See for instance: (Harding 1999: online; Kapustin 2000: online; Dobrovolsky 2009: online).

"We don't go out and seek our clients: *they come to us*, our clients approach us to provide them a service which they need," says Alesha, a gangster interviewed by Ries (2002: 280). In the same way, gangsters allegedly do not torment *babushki* who trade in the streets. The women themselves come to them and ask for protection from petty thieves and drunkards. "The traders and speculators are always cheating each other," the gangster explains, and in another place he goes on to say, "We protect our clients from being cheated, we enforce their contracts, collect their debts for them, ensure some kind of normal business relations. They know they can count on us" (Ries 2002: 279–280). In other words, entrepreneurs are protected from themselves. Policemen know about Alesha's gang but accept it, because they are not able to maintain order only by themselves. What is more, Alesha's gangsters are family men and decent people. They are disciplined: they are not allowed to drink too much and even to smoke. In other words, they obey certain rules, rules that bring to mind the thieves' code. That is why they are not just bandits, they are *honest bandits*. So is probably Tolya Bykov.

Another meaning of the term *honest* comes probably not from the entrepreneurs themselves. Already in socialist times some *vory v zakone* were perceived as Robin Hood figure, a good robber. One of them, a famous thief called Anatoly Cherkasov, taught his fellows to "take from those who have". This was in fact the case, for after all, *vory* extorted money from *tsekhoviki*, i.e. from the people who appropriated public property (Gurov 1990: online).

5. Embedding economy at Russian open-air markets

A Russian woman came to the stall and asked for 300 grams of dried plums. Feyrudin (T38) took a small shovel full of plums, put them into a plastic bag, and laid it on the electronic scale. There was a little too much in the bag, so he took back plums one by one until the scale showed exactly 300 grams. *It's ready, 45 rubles, please!*, he said to the lady. *You could have left this last plum*, noted the customer; she was disappointed that Feyrudin did not add the surplus as a bonus. *You Russians don't mind if I weigh too much, but if I give the exact amount, you're upset. I don't want your money but I don't present you with what's mine, either!*, Feyrudin replied quickly. *Don't say anything about the Russians!*, the woman shouted angrily and went away without taking the plums.

This argument was evoked by a plum. In a sense Feyrudin was a stranger inasmuch as he came from Tajikistan, but he was by no means a newcomer. He had lived in Russia for over a decade, spoke Russian fluently, had married a Russian woman, and had Russian citizenship. His high acculturation level did not, however, prevent this situation from happening, proving that at a Russian bazaar, ethnicity may matter in economic relationships. For instance, it can spring up all of the sudden, spoiling a transaction as simple as buying some plums.

The aim of this chapter is to investigate the factors that shape the economic behavior of petty traders. I will discuss particularly the different kinds of relationships between tradespeople, that is, different forms of cooperation and competition. The claim I make is that the economic behavior of traders is embedded in non-economic factors, such as ethnicity, social obligations, religious beliefs, and moral values. I follow here those scholars who argued that the cultural environment as well as personal ties and values have an impact on the way individuals manage their resources (cf. Bruni 2002; Granovetter 1973; Granovetter 1985; Polanyi 1957; Putnam et al. 1993: 170–172). In general, tradespeople are guided not only by pure calculation and individual economic rationality, which suggests they should just seek the maximization of personal profit. Especially in survival situations, their strategies of behavior are not simply of pure economic character but also include alternative rationalities of action.

In this chapter, the embeddedness of economic behavior of tradespeople in ethnicity is broadly examined. The question of ethnic economy or ethnic entrepreneurship has been widely discussed (Aldrich and Waldinger 1990; Bonacich 1973; Zhou 2004). It intersects with the phenomenon of trade minorities, which can be defined as ethnic or religious groups that specialize in market activity, first of all in trade (cf. Dyatlov 1996a: 318; Dyatlov 2000: 12). The problem can be expressed as follows: do migration and ethnicity constitute factors that influence the economic behavior of petty traders?

One has to remember that there is no long entrepreneurial tradition in Russia, neither an ethnic nor a general one. The last time that individual entrepreneurship was to a large extent allowed – and it proliferated in fact – was the era of the New Economic Policy, i.e. in the 1920s. Moreover, the most non-Russian traders came to Russia only after the dissolution of the USSR. By no means, then, do they form well-established trade minorities with centuries of entrepreneurial experience, like the Chinese in South-East Asia or Indians in tropical Africa (cf. Dyatlov 1996b). It is also because of these facts that some scholars claim that no ethnic-driven economy can be observed at Russian open-air markets. According to Ol'ga Brednikova and Oleg Pachenkov, migrant petty traders are guided solely by economic profit and follow free market competition rules: "Ethnic solidarity is limited by the notions of rationality" (2002: 157).

Contrary to those views, I argue – and this will be my first subclaim – that one can indeed speak about an ethnic economy and trade minorities in Russia. I follow here the position held by scholars such as Viktor Dyatlov (1996a; 2000; 2009), Dyatlov and Roman Kuznetsov (2004), Dyatlov and Konstantin Grigorichev (2013), and Pavel Varnavsky (2013). In practice, for instance, the eagerness to make individual profit is limited by the communities to which one belongs. These include communities that are spontaneously created by the traders at particular markets. In the case of immigrants, the most likely to appear is a closed community based on common ethnicity. Traders who have settled at an open-air market call their relatives and friends to come and thus a small trade minority might be created. However, it may happen that after people have gotten to know each other well and begun to cope with common challenges together, that transnational communities also appear. In both cases, individuals in their economic behavior often try to reconcile both their own and communitarian goals.

In their work and everyday life, petty traders usually have to cope with multiple problems: a severe climate, competition from supermarket chains and other traders, legislative regulations, and so on. Under such circumstances, many of them prefer to invest in social rather than purely economic capital. Investing in social capital means creating networks of trust. A trader at an open-air market needs people he or she can count on. As a matter of fact, it is quite difficult to work solely on one's own following purely individualistic principles. A reliable partner, a friendly neighbor, an honest supplier, or a decent seller whom you can employ are valued very highly among the tradespeople. Thus, the second subclaim I put forward is that building ties based on trust and forming communities are factors that support an individual, especially an entrepreneur, in coping with particular risks and uncertainties (cf. Putnam et al. 1993: 170; Sztompka 2003: 16–21). In this sense, the embedded economy reduces the extent of uncertainty a tradesperson has to cope with.

This does not mean that competition does not exist. Neither does it mean that there are no tensions between local sellers and people who come to work at Krasnoyarsk's markets. These phenomena certainly occur. Traders at an outdoor market constitute a collection of very different people of often contradictory interests, so conflicts between them are likely to appear. They are of various origins and ethnic background, including both local people, mainly Russians, and labor migrants who have come to Russia from the former Soviet republics. Citizens of other countries, such as China or Vietnam, also come to some of the markets. Moreover, traders are of different economic status; most of them are individual entrepreneurs running their own companies, but some are hired sellers. The amount of turnover and profit made by entrepreneurs can vary significantly as well. Last but not least, frictions arise due to the simple fact that traders compete with each other for clients. Nevertheless, there are also aspects that unite all tradespeople who work at Russian open-air markets. Competition – like other forms of economic behavior – is embedded, too.

I begin the chapter with an account of the history of the Slavyansky open-air market in Krasnoyarsk. As a result of legislative regulations and market processes it was closed in January 2014 after years of fighting for its existence. I treat the Slavyansky's case as symptomatic for the evolution that Russian petty trade has undergone since it erupted in the 1990s. Next, I recall two stories of tradespeople from the Slavyansky: Tanya and Fedya. The role of the stories as well as of the accompanying comments from my other interview partners is to show the conditions of working at Siberian open-air markets, the values attached by tradespeople to their labor, and the different modes of running a business. In addition, the fieldwork material shall provide an insight into the complex issue of relationships among tradespeople themselves and between tradespeople and their customers. In some aspects, the stories may contradict each other, since my intention is not to deliver a coherent picture of Krasnoyarsk's outdoor markets – that would be impossible – but to give a general idea of the possible forms that social relationships at a market can take on. In the following sections, I discuss the issues of ethnic economy and trade minorities. I show how non-economic factors, first of all ethnicity and communality, may influence the economic behavior of traders.

Living in legal uncertainty: the Slavyansky market epic

The Slavyansky in the Zheleznodorozhny district of Krasnoyarsk was one of the oldest open-air markets in the city. It appeared around 1995 as a *stikhiyny rynok*. In February 2005, a company called Slavyansky Bazar-Kredo-7 (LLC) took out from the city a lease on the place in order to organize an outdoor market there. Hence, the market had already been operating for a decade when it was officially founded. Formally, the agreement with the city was valid until the end of the year, but the Slavyansky continued to operate afterwards. It was only in March 2007 that the next contract was signed, which, in fact, was a prolongation of the 2005 agreement. However, this latter document was in effect for just a few months and ceased to be valid in the beginning of August 2007.

In July 2009 a decision was made in the Krasnoyarsk administration to build in place of the open-air market a closed one. It would be a trade center with underground parking. The company Sirius (LLC) was chosen to carry out that task. A year after, in June 2010, the city signed an agreement with the company and the latter got a lease for the building plot for three years. One could have thought that the open-air market was already doomed. Nevertheless, only in February 2014 did Sirius start to clear the site of the former market in order to prepare the construction of the new trade center.

The main reason for building a trade center in place of the open-air market were legislative changes. As I wrote in the Introduction, the new act *On retail markets and on introducing changes to the employment code of the Russian Federation* was passed by the State Duma and eventually, after much discussion and revision was signed by President Medvedev and came into force on January 1, 2013. The act led to closure or thorough reconstruction of markets in entire Russia. This process started already in 2006, when the new law was proposed.

There are two meaningful sayings which describe how laws are implemented in Russia. The first says that *the strictness of laws in Russia is lessened by their non-obligatory implementation* (*strogost' zakonov w Rossii smyagchayetsya neobyazatel'nost'yu ikh ispolneniya*).[42] The second proverb is that *the law is like the steering of a horse carriage:*

42 This saying, recalled in various forms, is often attributed to the famous Russian writer and satirist Mikhail Saltykov-Shchedrin, although one cannot find it in his writings. One of the sources may be traced in the works of the poet Pyotr Vyazemsky who writes,

wherever you direct it, there it leads you (*zakon eto dyshlo: kuda povernul, tuda i vyshlo*). The latter saying, as one of my interlocutors explained to me, perfectly describes the selective way with which Russian prosecutors' offices apply the law. In the case of the law *On retail markets* and the Slavyansky, both of the sayings seem to apply. Since 2007, when the first rumors about its closing had spread, the market continued to operate for over 7 years.

Kredo-7, the market's administrator, gained some time by bringing a few lawsuits against the city administration. It tried to prove that the lease agreement from 2005 was still valid, and thus that the agreement the city signed with Sirius in 2010 was unlawful. In July 2012, however, the arbitral tribunal of Krasnoyarsk pronounced its sentence: Kredo-7 had no right to occupy the site of the market and the market itself was illegal.

During the judicial proceedings, a few other things were proven: the city administration had issued no permissions for the functioning of the Slavyansky since August 2007. Therefore, all the subleases Kredo-7 had been signing with tradespeople – that is, contracts which allowed particular entrepreneurs to rent stalls for trading at the market – were illicit. Kredo-7 simply had no rights to the market plot, so it was not in a position to draw up any agreements concerning it. In addition, the area of the market did not correspond to the urban land development plan. There had also been no permissions issued for building *temporary structures* there since 2005. Such structures were nevertheless built. Last but not least, the market did not fulfill the sanitation requirements for carrying out retail trade.

Surprisingly enough, notwithstanding all the suits Kredo-7 filed against the city administration, the authorities of the Zheleznodoroznhy district allowed the company to organize a fair in the contentious area. Calling the place a *fair* (*yarmarka*) and not a *market* (*rynok*) was a legal trick. The 2006 law concerned retail markets and even clearly stated that it did not regulate the functioning of fairs, i.e. trading festivities organized occasionally on holidays. The Slavyansky fair was not occasional at all: it worked 7 days per week and it was just the name that differentiated it from a regular market. The permission from the authorities of the Zheleznodoroznhy district was issued for just 3 months from January to March 2012. For whatever reason, Kredo-7 did not even apply for a prolongation of the license. From March 2012 until its very end in January 2014, the Slavyansky worked as before, being again of illegitimate status.

In summary, since its actual beginning in 1995, the market was legal in three periods. The first one was from February 2005 until the end of that year. The second period lasted from March to August 2007, though during this time the market's legality was also extended *a posteriori* to include the period from the beginning of 2006. Finally, the third period lasted from January to March 2012 when it was registered as a fair. Thus, throughout the majority of its history, the Slavyansky was in fact illegal. How it persisted for so long in spite of this fact remains a mystery.

Indeed, as stated above, Kredo-7 successfully bought time by bringing lawsuits against the city administration. These trials, however, cover just a small portion of the Slavyansky's period of illicit operation. As a matter of fact, Sirius – the developer

---

"I think it was probably [Grigory] Poletika [another Russian man of letters] who said: there is a rescue from idiotic measures introduced by the government in Russia: their idiotic implementation" (Vyazemsky 1963 [1878–1888]: 24).

of the new trade center – could have started construction, for instance, in June 2010, when it signed the contract with the city or in July 2012, when the city won an important lawsuit against Kredo-7. In other words, the story of the Slavyansky could have ended drastically with bulldozers and bailiffs coming to the fully operating open-air market. However, this did not happen. For whatever reason, Sirius was very patient in executing its rights and waited for years until the Slavyansky finished its existence in a peaceful way.

This patience on the part of both Sirius and the city is particularly astonishing when it comes to the most mysterious of all things: money. Tatyana Dal' from the Krasnoyarsk municipality stated in a TV interview that since August 2007 the city's budget had not received even a kopeck from the administration of the Slavyansky. She accused the managers of Kredo-7 of taking all of the money from the tradespeople's rents to line their own pockets (Istomina 2014: online). The question of where the money from the Slavyansky traders went was frequently asked while assessing the market's status. However, one of the arguments Kredo-7 used in 2010 and in the following years to support its claim about the validity of the 2005 agreement was the fact that it regularly paid urban taxes. Eventually, the court ruled the Slavyansky to be illegal not because Kredo-7 had not paid the rent but because its agreement with the city was no longer valid. Although some concerns about the social role of the market might have toned down the city's policy towards it, one finds it hard to imagine that urban officials would just let a private company make a profit for 7 years without paying any taxes.

To make the issue even more confusing, Sirius representative Vasily Yeltishev insisted that since 2010 it was his company which paid the rent for its potential construction site. He specified that the monthly rent was 100,000 rubles, which added up to a considerable sum of 1.2 million rubles (approx. 40,000 US dollars) per year.[43]

If the words of Yeltishev were true, at the time he spoke Sirius had been paying high rents for the area for three years already. What is more, the company had been doing it while patiently waiting for tradespeople from the Slavyansky to leave their workplace peacefully. The only explanation for such behavior and, in general, for such a long story of the market's illegal operation is that there were some informal agreements, presumably financial in character, between the main actors: Kredo-7, Sirius and the city's administration. Otherwise how could a *de jure* illegal market continue operating for almost 7 years?

Coping with uncertainty in workplace: the Slavyansky traders

What happened to and around the Slavyansky is more or less clear now and – except for a few aspects that remain mysterious – one can reconstruct it from court decisions and countless press reports (cf. Aleksandrov 2013: online; Arbitraznhny sud Krasnoyarskogo kraya 2013: online, Borkozhokova 2013: online; Istomina 2014: online; Semikina 2013: online). Nonetheless, while the events were taking place, the tradespeople hardly knew what was happening with their market. When I came to the market for

43 Yeltishev stated this during a public meeting concerning the situation of the Slavyansky on September 13, 2013. After the meeting, he personally confirmed to me this fact.

the first time in February 2012, different stories circulated among the tradespeople, but nobody knew anything for sure.

What people did know was that their market was deemed to be closed. The first rumors about it appeared already in 2007. However, years went by and nothing had really been changing. Only in 2010 did people realize that the situation was getting worse: their permissions for trade, at first issued for 6 months, had to be renewed every quarter and eventually began to be issued on a month-to-month basis.

One of the pieces of gossip said the area of the market would serve as the place for a new housing estate. The Slavyansky was situated not far from the city center and thus constituted an attractive location for development. Right at its border a large apartment block was already being built and rumor had it that the developer had his eye on the market itself. People heard that the developer had sued the city, demanding a part of the market's area to locate facilities for the building. This was true. The lawsuit was, however, later dismissed by the Arbitral Tribunal of Krasnoyarsk. Yet in 2012, the danger looked real. The Slavyansky was already squeezed between multistory apartment blocks. In addition, the tradespeople knew the fate of the GorDK market.[44] Despite violent protests from people who had worked there, it had been closed and a new housing estate was being constructed in its place.

Nevertheless, in the case of the Slavyansky nothing was settled yet. "*The worst thing is the uncertainty*," reported Lyuba (T59), the tradeswoman whose story was told in Chapter 4. She opted out of buying a batch of goods at an attractive price. She did not want to risk not selling them during the next month. A month was the biggest time horizon she could plan and that was much too short for her to be able to run her enterprise in the usual regimen. The other traders at the Slavyansky were in the same situation. It does not make sense, for instance, to renovate your booth if you are not sure whether it will still be yours in the near future. Uncertainty afflicted the business activity of the Slavyansky vendors, making their various decisions difficult, but it had one common dimension: no one knew whether the market would really be shut down and what they were going to do if this happened.

Operating in a state of constant uncertainty of their future was just the tip of the iceberg for the traders. A few years before, the Slavyansky had been an attractive market for tradespeople, because its abundant customers offered a possibility to earn decent money. However, when I visited the market for the first time in February 2012 the situation was quite the opposite. Only some of the vendors still enjoyed a large demand, while many others were just trying to make ends meet. The reasons for such a turn of events and for the fact that many Slavyansky traders were forced to quit were basically twofold. Firstly, the global financial crisis had an impact on Russia, too. On the one hand, it caused an increase in foodstuff prices, while on the other, it diminished the purchase capacity of the average customer. The result was simple: people could buy less. Secondly, two supermarkets opened in the neighborhood of the market. They offered comparable prices and many customers preferred to do their shopping there, because – as tradespeople took it – the supermarkets were clean, more civilized, and heated during the winter. *This market is dying* was one of the first sentences I heard at the Slavyansky.

---

44 The abbreviation *GorDK* comes from *Gorodskoy Dom Kul'tury* (*Municipal Cultural Centre*). The open-air market lay just a few bus stops from the Slavyansky market. After the closure of the GorDK some of its traders moved to the Slavyansky.

Still, people seemed to be working as usual – some of them barely sustaining their existence, some earning on average, but some – in spite of all difficulties – doing relatively well. What united all of them was holding tightly to their workplaces. "*We are sitting on a powder keg. I do not wish my worst enemy to be without a job,*" said a tradeswoman in a TV report from the market (Bugakova 2013: online). "*The market's future is up in the air,*" Feyrudin (T38) told me when I asked him what would happen to the market. According to the data I acquired from the administration of the Zheleznodorozhny district, in September 2013 there were as many as 243 entrepreneurs registered at the Slavyansky. Since some of them hired sellers to work at their stands, one can assume that the market provided support for around 300 households. Because the traders were responsible for the fate of their families, they tried to stay at the Slavyansky as long as they could.

The entrepreneurs had been trying to defend their workplaces since 2010. Several petitions were signed and taken to the urban authorities. As I was told, an open letter was even sent to President Putin himself. Interestingly, the letter was written on behalf of not only the traders but the local population as well. A wide range of products could be acquired at the market, ranging various different foodstuffs to household items and clothes. There was also a separate part consisting of so-called social spots, where pensioners could sell their produce without paying any rent. Thus, the market gained considerable social importance and there were many people who had simply got used to shopping there and joined in the attempts to save the place. Putin was said to have answered that he left the solution of the problem to the municipality.

Additionally, the Slavyansky's case was also picked up by some politicians during the campaign before local elections in September 2013. One of the rumors had it that the market was under the protection of Tolya Bykov, the politician described in Chapter 4. Because of Bykov's *krysha* – the story went – the Slavyansky was less harassed by racketeers in the 1990s and now it would prevent it from being shut down.

Although pre-election promises raised some hopes among some of the tradespeople, in general they did not believe in politicians at all. *They do not give a damn about us* (*im na nas naplevat'*) said Ibrahim (T17). This might serve as a quintessence of the traders' attitude towards the authorities and all people from privileged classes. The traders did not expect any support from them, since their goal was not to benefit the people (*narod*) but to acquire more wealth and power. In addition, nobody at the market considered an active protest in the form of a demonstration. After the rallies in Moscow's Bolotnaya Square in 2011–2012, unregistered protests were brutally dispelled by the police and nobody was eager to take that risk.

In the second half of 2013, a billboard was put up in the neighborhood of the market. It depicted a future trade center. This way the tradespeople knew at last what was going to happen to their market. The question they faced, however, remained the same: how much longer would they be allowed to work? During a public meeting in September, it was negotiated that they had until the end of the year to close down their activities.[45] It was also proposed that they be able to move to other urban markets.

45 The meeting, as was mentioned above, took place in the seat of the administration of the Zheleznodorozhny district on September 13, 2013. Its participants included a representative of the Sirius company, Vasily Yeltishev; district officials; and a handful of entrepreneurs. The market administrator Kredo-7 was not represented. A telling detail was that the traders were only informed about it the previous evening. As a matter of

The possibility to move their businesses to other markets was something the vendors had already heard before, but hardly anyone considered it an attractive option. What the people from the Slavyansky and from other markets often stressed is that they lived first of all on steady customers. Gaining such customers is for petty vendors one of the best economic strategies. However, while a steady customer might choose to buy sausages at your stand and not a neighboring one, he will not follow you to another part of town. In other words, location matters for both vendors and their clients. Rarely does one buy basic foodstuffs in a shop distant from one's home. Thus, moving to another market results for a trader in cutting all the ties he or she has managed to build with customers. Moreover, a new market already has its incumbent tradespeople; it is very difficult to open a new stand at a market and compete with vendors who have been working there for a longer time.

At the meeting in the district administration office there was also a lot of deliberation as to whether the traders might find a place in the new trade center, once it was build. Earlier, some city officials and journalists assured that the rent prices would be the same as at the market (Orlov 2013: 4). None of the traders believed that. At the meeting, the developer neither confirmed it nor guaranteed that there would be a space for individual entrepreneurs at all. It was declared that space in the new center would be leased according to "market laws", which meant rental fees would be set by the owner of the center. It is, however, obvious that for a building owner it is more convenient to lease an entire floor to a supermarket than to divide it into smaller parts, to sign several agreements with individual entrepreneurs, and to fulfill the specific needs of each particular shop. Moreover, in competition between retail networks and petty traders it is the former that usually win. Networks simply have much more financial sources at their disposal, so they are better suited to fit so-called market laws whatever they might be. Thus, the Slavyansky entrepreneurs basically lacked any certainty whether they would ever be able to come back to the place.

Although theoretically the Slavyansky tradespeople had until the end of 2013 to wrap up their businesses, in October of that year the power supply to the market was cut off. Apparently, the administrating company accepted its failure and stopped paying their bills. Winter was mild that year, but it was still extremely difficult to work at the market without electricity. While some traders tried to produce it on their own, bringing fuel generators to provide some light and heating to their booths, most were forced to leave. In the last days of the Slavyansky at the end of January 2014 around 70 entrepreneurs still continued to work there. The last ones were only expelled from the market by bailiffs (Istomina 2014: online; Polyakov 2014: online).

❖ *It is better to have 100 friends than 100 rubles:* the story of Tanya

Tanya (T4), a Russian woman with Polish roots, worked at the Slavyansky for over 10 years beginning from 1996. This was the time when the market flourished. At the beginning she worked as a seller for another person. She was cheated, however, as her employer did not pay her salary. At that time, she said, it was a common practice for the owners of the stalls to hire young women as sellers and often they would

fact, they were invited just to be notified about the necessity that they leave the market, not to be heard. However, the head of the district administration Yury Savchuk played the role of the traders' advocate.

exploit them by just refusing to pay their wages. Therefore, Tanya decided to work on her own, thus becoming an individual entrepreneur. She started to sell newspapers, journals, crossword magazines, and so on.

The work was hard. Since Tanya lived 40 kilometers outside Krasnoyarsk, she had to wake up early in the morning, go to the city by train, collect the newspapers from the printing house and only then come to the market. Working at an outdoor market has a specific rhythm to it: the best days are weekends and holidays. Obviously, customers have more time to do their shopping then. The only day off for a petty trader is usually Monday. Of course, in the case of entrepreneurs it is they themselves who decide how much they work. Since earning one's living from petty trade is challenging today, they often decide to work as much as possible. So, it was only on Mondays that Tanya stayed home, took care of the household, and did her *female stuff*, i.e. tiding up or cooking.

Particularly difficult were winters. In the winter time, which in Krasnoyarsk can last for up to 6 months, temperatures can fall to -40° Celsius. It is only in recent years that outdoor markets in the city, and in Russia in general, are being renovated, with common roofs being built and people turning their stands into small kiosks. In the 1990s, however, literally everybody worked outside, even through the winter. For many people this is still the case. Tanya had only a small stand and the sole thing she possessed was a tiny roof more or less protecting her goods from rain. She said she would not work only if the temperature was below -30°. From my own observations I can state that when the temperature drops to -25°, markets die out, as customers are simply not very eager to do their shopping in such cold weather.

During her first months, Tanya did not do well. She had to learn to do business. An additional factor was the competition. There was another woman at the market who traded in newspapers as well and offered attractive prices. In Krasnoyarsk, employees of the printing house worked on the side and what they printed overtime, they sold directly to stand holders. Tanya's rival got newspapers from them and Tanya, who acquired her goods via official distributors, could not compete with her. It was only after she made her own informal connections that good times started for her.

What Tanya enjoyed the most was independence. As she stressed, it was *psychologically* difficult for her to subordinate herself to somebody. She worked a lot, but it was of her own will. She could decide about what goods to sell, too. In the socialist era she had worked in a public library. She had to start working at the market after her position was cut. However, selling newspapers seemed for her a kind of continuation of her previous work. In addition, she occasionally supplemented newspapers with seasonal goods like grains or plants, because, as she put it, one cannot survive by selling only one type of merchandise. She also enjoyed interactions with the customers. With some of them she established deeper relationships and sometimes they would come and tell her about their problems. Perhaps all these factors contributed to Tanya's success: with her earnings she managed to buy a house. The house was not in Krasnoyarsk itself but in her home town and required a complete renovation, but it was a whole house.

Tanya's attitude to labor migrants coming to Krasnoyarsk was rather unemotional. She observed with a degree of concern that migrants slowly drove Russians out of petty trade. She also expressed the usual complaint that newcomers spoil the labor market because they accept much lower wages than local inhabitants would. However, she had good relationships with foreigners, too. An Uzbek man named Nurdzhigit became one of her best friends at the market.

The 1990s and the early 2000s were a tough time for labor migrants in Russia. They were an attractive target for various law-enforcement bodies, which fined them on every possible occasion. Outdoor markets were particularly kept under close tabs. There were – and still are – cases in which special OMON police units would close every exit from a market and detain every labor migrant they could find. Nurdzhigit was safe, as he had married a Russian woman and had Russian citizenship. Paradoxically, it was Tanya who was in trouble because one day her papers had been stolen. There was a police officer who constantly pestered her. Obviously, he was expecting a bribe, but Tanya did not want to hand it over to him. It was Nurdzihigit who used to hide her in his container until she finally procured new documents for herself. He helped her in other ways as well, heating weights on an electric cooker so she could warm herself up holding them. Tanya was thankful. When Nurdzhigit was later forced to leave the market, she sold his goods for him. All in all, Tanya managed to establish good contacts with many of the Slavyansky tradespeople.

As I observed at the Slavyansky, people would work on average 8–10 hours per day, 6 or even 7 days a week. It is simply much better for a trader to have good relations with those he or she spent so much time with. It is also because of this circumstance that people at the market form specific communities. They jointly celebrate birthdays and other holidays, such as Women's Day. Usually neighbors put a table in a common place so that while eating, each of them can have an occasional look at his or her stall. They also support each other: when the husband of one of the tradeswomen died, other traders collected money for her, and even those who hardly knew her contributed to this assistance.

Yekaterina (T6), a friend of Tanya, did not enjoy such support when her mother died. The only person she could ask for help with the funeral was Tanya. In the latter's opinion, this was due to the fact that Yekaterina focused too much on her job, which prevented her from building deeper relationships with other traders. In other words, Yekaterina invested in economic rather than social capital. Meanwhile, as Tanya said, *it is better to have 100 friends than 100 rubles.* This powerful expression constitutes an important part of the common knowledge of every Russian and is quite often used by petty traders. Tanya herself was an example of successfully combining investment in economic as well in social capital.

❖ Achieving economic goals with magical means: the story of Fedya

Fedya (T9) was 43 years old when I met him in February 2012. His actual name was Farrukh, but he introduced himself as Fedya and everybody called him by this typical Russian name. Though he came from Azerbaijan, he had been living in Russia for 25 years, had married a Russian woman, and was a Russian citizen. Earlier, he lived in Norilsk in the Russian Far North. He worked as a plumber and a welder there, but he was dismissed from his job on disciplinary grounds, for he drank at work. Therefore, he began to trade. Then, he and his wife moved to Krasnoyarsk.

Fedya started to trade at the Slavyansky in 2008. The reason he chose the market was that he lived in close vicinity to it. He sold fish and his business was developing. In addition to buying a double place for himself, he managed to buy another place at the market as well, turn it into an enclosed shop, and fully equip it. His wife worked there. As time went by, however, his profits began to constantly decline and his business did worse and worse. When we met he was on the verge of bankruptcy. To pay the rent for April 2012, for instance, he was forced to borrow some money from his friends, other Azerbaijani traders. He worried about the whole situation and he said he was constantly thinking about the problems he was facing, which is why he could not sleep. In general, Fedya suffered from depression.

Fedya considered employment at the market not to be true work. For him, true work was when you work in a factory and get paid every month. He also said that the life of a trader is not a human one but a dog's life. Soviet times had been good, but since the collapse of communism the quality of life had been getting worse and worse due to a whole set of reasons, such as poverty and misfortune, environment pollution, and artificial foodstuffs produced by the Chinese. Russians in his view had become evil and corrupted. "*If I fainted in the middle of the market and fell on the ground, nobody would come and pick me up*," he said. He claimed that Russians – because of their allegedly diminished morality – were no longer human. "*They are shells, they only look like people but inside they have become rotten*," he kept repeating.

He despised Russians and at the same time he felt deep resentment towards them for not acknowledging him as one of them. "*You and I – we will never feel at home here*," he told me. The difference was that I had no intention to become a Russian. He, on the contrary, had been living in Russia for over 20 years and was a Russian citizen, but he was not accepted as such by the local population. Nevertheless, he despised labor migrants as well. *Pizdoglazye* (*the cunt-eyed*) was the word he used to describe the Kyrgyz and the Chinese. He was neither a local nor a migrant. Rather, he was in-between.

Fedya saw the causes of his business failure not only in the financial crisis and competition posed by two supermarkets which opened in the neighborhood. He saw it also in the magical assaults that – as he claimed – had been carried out against him by other traders from the market. The reason why they did so was that in 2010 – that is, 2 years after he started to work at the Slavyansky – he began also to sell fruit at his place. This caused a conflict with his neighbor, a Russian woman called Lena, who also sold fruit. Lena went to a female sorcerer, a so-called *babka*, and received some evil spells from her. Then she started to cast them on Fedya.

The spells were successful. Fedya's business stopped bringing profits, because even his regular customers ceased to buy from him. Moreover, Fedya suddenly got ill. He had problems breathing and was so weak that he could not raise his hand. He understood that something was wrong, so he also went to a *babka*, albeit a different one. This sorcerer gave him a magical amulet and said it would deflect the spells. Fedya showed me this amulet, which he wore around his neck, but I was not supposed to touch it. The amulet was round in shape and a sun was depicted on it.

The amulet really worked. The spells cast by Lena were returned to her. As a matter of fact, it was only this way that Fedya recognized who had attacked him. The woman got an allergy on her cheeks and sometimes when looking at Fedya she would even faint. Fedya himself, in turn, recovered quickly and when we met he no longer had problems with his health.

Nonetheless, the amulet did not help bring back his customers, so his business did not improve. Thus, he consulted another *babka* to find a solution. The new *babka* explained to him that Lena had allied herself with some other traders from the market.

These other people traded with fish, like Fedya. Because of the crisis they, too, had financial problems and thus were interested in causing him trouble. "*It is like you are in a bucket*," said the *babka* to Fedya, which meant he was surrounded by enemies who constantly kept attacking him with evil spells. Fedya also told me he would often find salt spilled round his stall. It was his enemies who charged the salt with evil energy and spilled it to draw customers away from his stand.

The *babka* explained to Fedya what was happening, but she could not help him even though he gave her a handsome sum of money. She might have been not skilled enough to defend Fedya, but in fact Fedya thought that she did not want to help him because she was Russian like his enemies. That is, she would not harm her own people. Whether it was because of the magical assaults or not, eventually Fedya was forced to close his stall and, for a period of time, leave the market.

Magic was a way in which Fedya tried to save his business. Galina Lindquist in her study on magic and healing in contemporary Russia wrote: "Magic practices thrive where power is brutal and overwhelming, where the rational channels of agency are insufficient or of limited value, and where the uncertainty of life calls for methods of existential reassurance and control that rational and technical means cannot offer" (2006: 2). In general, people use magic "when other channels of agency are blocked" (Lindquist 2006: 9). Turning to an occultist then often means that a person is utterly cornered, as for many people it is something shameful and stigmatizing. Magic is often the last resort for people who are too weak to act in usual ways.

This was the situation of Fedya. He fought for survival. The usual means of running his business – such as good relationships with customers and offering them high quality goods – proved insufficient. He got sick and everything he had collapsed into ruin. Thus, he had to find an alternative way out of his situation. If one understands rationality prosaically as goal-oriented and well-planned action (cf. Jarvie and Agassi 2010: 593), Fedya's strategy was rational. This rationality was even confirmed by the results, as Fedya recovered and it was his enemy, in turn, who got sick.

Yet, it was a rationality other than the economic one. On the one hand, survival is something which belongs to the sphere of economy: you have to feed your children, you have to pay the rent for the apartment, and so on. On the other hand, according to economic rationality turning to a *babka* would be a pure waste of money. Thus, one can say that Fedya's survival strategy was economic even though it had no economic sense. It was an economic behavior embedded in magical, or religious, beliefs.

There are some practices at Russian open-air markets – besides the one described above – that could be described as superstitious. You should, for instance, brush your goods with the money you received from the first customer of the day. You should also give the first customer a small discount, especially when the customer is male, because a female one brings rather bad luck. If people perform these acts, they do so with some indulgence, though. After all, brushing goods with money will not disturb your business and if it might help, why should you not do it? Many traders would agree that religious or quasi-religious factors might affect your performance at a market.

For instance, I heard opinions from vendors that people are charged with certain energies. If a person you have encountered possesses a positive energy – which can translate to a good mood, positive thinking, nice words, and so on – meeting with this person will somewhat charge you up. On the contrary, if the person has a bad energy, the meeting will make you down and you will feel weaker after it. Sulayman (T30), who worked at the Slavyansky, told me that the first thing he does after coming back from the market is to take a shower. He does so not because of any material dirt but in order to clean himself from the evil emotions he was exposed to during the day. He said that the customers' greedy looks or your market rivals' bad wishes might burden you or even make you sick, physically or mentally. Sulayman did not confine himself to taking showers, for as a devout Muslim he prayed every day for protection from all evil forces that could harm him.

I was told a similar thing by Kamran (T20). He said that out of misery and poverty, many people come to the market just to express their anger. They approach your stall, buy nothing, but only complain about the quality of your goods or your prices. I observed such cases many times. With his brother Ibrahim (T17), Kamran tried to completely ignore such a person so as to not be thrown off balance and to make the person leave as quickly as possible. Nonetheless, positive situations happened to them as well. Once, when I visited the stand of the two brothers and was helping them as usual, all of a sudden customers started to come in numbers. They did substantial shopping and each of us had his hands full serving them. When this unexpected wave passed, Kamran commented on it, saying I had a *light hand.* It was I who – in his opinion – had attracted the customers. The *light hand* meant I made the goods sold by the brothers leave their stand easily. It was as if I had brought a positive energy to the place.

Fedya's case shows that in survival situations people turn to strategies which are alternative to purely economic ones. Trade in fruit and vegetables at the Slavyansky was particularly competitive. Previously it was almost entirely Azerbaijanis who dealt with it. After the liquidation of the GorDK market, however, fruit and vegetables sellers were among the traders from there who came to the Slavyansky. This was the case of Sergey (T26) and his wife Lena, who later became Fedya's main enemy. In addition, the outflow of customers from the market made some of the Kyrgyz give up trading in clothes and switch to fruit and vegetables. As a result, there were 8 stands around a small square at the market lying practically next to each other and offering the same goods. At one point the unstable balance was disturbed and some of the traders started to seriously compete with each other.

Thus, maybe there was nothing extraordinary about Fedya's failure. Ibrahim and Kamran, who were Fedya's neighbors at the market (from another side than Sergey and Lena) had a bad opinion of him. As a matter of fact, they were rather ashamed that he is Azerbaijani, as well. They condemned him for being rude, untidy, and drinking too much. All of that was true: Fedya cursed a lot and did not care about his appearance. Moreover, he sometimes drank vodka even during working hour, even though tradespeople – as I was explained – virtually never do it. Apart from many other negative effects, it makes a very bad impression on customers when you smell of alcohol.

I did not manage to talk to Lena to prove Fedya's interpretation. This was probably due to the fact she saw me several times talking to him. Whether or not because of Fedya's magical counterattack, she indeed suffered from an allergy on her cheeks. I frequently talked to her husband, Sergey, but we did not touch on the issue of magic. She has never participated in these talks, however. She sat aside, trying to not even look at me. Her strange behavior made me think that indeed there was something unusual at stake in this competition. I also saw Fedya's amulet and – above all – why should I not believe him? Apparently, "in the world where magic is practiced, it does indeed work" (Lindquist 2006: 4).

After he quit trading fish in April 2012, Fedya performed some chaotic moves. He had already had debt, but he somehow managed to take out more credit, being in fact hardly able to pay it off. He joined forces with Alik (T12) and together, they took up trade in fruit and vegetables. When this did not succeed, he started to cooperate with an Azerbaijani tradeswoman. He bought the goods – biscuits this time – and she sold them at the stand. They were supposed to share the profits fifty-fifty but again, the plan misfired within a few weeks. Eventually, Fedya was forced to leave both his double stand and his fully equipped shop. He tried to sell the places, but since the Slavyansky's fate was uncertain there were not many potential buyers. His place was finally taken over by his enemies – Sergey and Lena. When I came back to Krasnoyarsk a year later in August 2013, he was helping his friend in a stand with fruit and vegetables outside the Slavyansky. He made plans for moving to another city, since Krasnoyarsk was clearly a place which did not bring him luck.

Ethnic economy and trade minorities

Like other Russian regional centers, Krasnoyarsk is among the destinations of working migrants from the so-called near abroad (*blizhneye zarubezh'ye*), i.e. the former Soviet republics. It is especially people from Central Asia and the Caucasus who massively come to the city in search of employment. Significant immigration from China also takes place. The large inflow of migrants leads to the rise of new divisions not only in the labor market but in urban geography as well.

The Tajiks work on construction sites, the Armenians build roads, the Kyrgyz are housekeepers, the Chechens run restaurants, the Azerbaijanis and the Chinese work at outdoor markets – such classifications are easy to hear and form a kind of commonplace knowledge in Krasnoyarsk. For instance, according to Aleksandra, a Russian saleswoman, *all Uzbeks drive public buses; drivers and conductors – they all are Uzbeks* (T11). It seems that entire professions have become monopolized by members of a single nation. The same process can be observed in the area of petty trade, too.

As I suggested in the introduction to this chapter, I argue that one can observe phenomena tied to ethnic economy and trade minorities in the realm of petty trade in contemporary Russia. Such economy is characteristic of professions of lower status: labor migrants usually perform jobs that the local population take unwillingly, at least for the wages that migrants accept. As was said above, newcomers work on construction sites, as bus drivers and as housekeepers, in petty trade, and so on. In colloquial Russian, expressions like *a Tajik labor, a Tajik salary* or *to work as a Tajik* refer to hard work for little money. Admittedly, such expressions are taken from an ethnic background but point more to general socioeconomic deprivation than just to Tajik people themselves (Dyatlov and Grigorichev 2013: 28).

Labor migrants often take jobs in petty trade, since it is a business relatively easy to enter into. You do not have to have special education; even proficiency in Russian is not necessary. You do not have to possess large starting capital, either. If you have no money at all, you can always find employment as a salesperson. On the other hand, in the Siberian climate, trading at an open-air market is a difficult way of life. *There are almost no Russians left at the markets* said Nadya (T5), a Russian tradeswoman from the Krastets market in Krasnoyarsk. This opinion, although it seems like an obvious exaggeration, expresses a clear tendency: at most outdoor markets in the city it is Russians who have become a minority. According to Nadya, one of the reasons for this is that Russians simply do not want to work in such conditions. As Nadya said, *I don't think you could force our youth today to do such work.*

Some ethnic-based divisions can be seen not only between various professions but within different branches of petty trade, too. For instance, at the Slavyansky market the Kyrgyz generally traded in clothes, Azerbaijanis in fruit and vegetables, and Russians in all other food products. These divisions translated also into space. While the Kyrgyz occupied the center of the market, Azerbaijanis concentrated rather by its southern gate and Russians worked along its perimeter. At the Krastets, the city's biggest open-air clothing market, on closer examination, one notices that shoes are sold mainly by Tajiks, jeans by Azerbaijanis, and jackets by Kyrgyz.

Such concentrations and specializations took place spontaneously. The crucial economic decisions of an individual – where to emigrate, where to work, and what to sell – are shaped by his or her personal networks. "*It was our neighbor Altay who came to Krasnoyarsk as the first one. Me and my cousin Alik joined him afterwards,*" I was told by Halay (B6), an Azerbaijani trader from the Yuzhny market. This is a typical situation: family and friendship ties constitute the channels through which a migrant comes to a foreign country, chooses a particular city, and takes a specific job. Friends and family members help the newcomer with getting set up – finding accommodation, getting work experience, legalizing one's residence, and so on. Newcomers engage in the branches that their friends, neighbors, and relatives had taken up before them. In such a way migrant communities grow bigger and bigger and become stronger, which in turn allows and forces them to compete with other minorities. Such communities are characterized by close ties. Basically, unless a market is a big one, all members of particular ethnic group know each other.

Indeed, one can ask whether the relationships between members of a trade minority are primarily of an ethnic character. This is a serious question in so far as it touches on issues of personal identity and the social roles an individual plays in his or her life. As Brednikova and Pachenkov rightfully note, one cannot disregard the fact that someone has family, is an alcoholic, or is unemployed and take into consideration just his national identity (2002: 160). Obviously enough, Kyrgyz labor migrants come to Russia not because they are Kyrgyz but out of economic necessity. However, their Kyrgyz background becomes important when they are abroad, often in surprising way.

Varnavsky, in his study of Kyrgyz labor migrants in Ulan-Ude (2013), showed that the ethnicity of the migrants actualizes upon their encounter with a strange environment. After coming to Russia, the migrants become Kyrgyz, even if earlier they might have identified themselves rather with family ties or another ethnic group living in Kyrgyzstan.

The common situation of being foreigners is something which unites people across divisions. For instance, it makes the strong regional differences between the northern and southern parts of Kyrgyzstan fade into the background. Varnavsky provides an example of a Kyrgyz woman who only in Russia learned to speak Kyrgyz. He describes also a process of forming the Kyrgyz diaspora in Ulan-Ude. From an accidental group of people who could count only on their families, Kyrgyz migrants had consolidated themselves into a dynamic community. An important role in this process was played by the Kyrgyz Cultural Center, an institution established by the migrants themselves. The Center now represents the community in relations with the authorities, supports newcomers, and organizes communal events, such as a football championship of Kyrgyz diasporas throughout Russia.

In other words, migrants who come to Russia become ethnicized. This ethnicization takes place on two levels. The first one can be called internal: migrants become aware of and embrace their ethnicity. The second level is external in so far as it is imposed on migrants in a way. It is the local population and especially the local mass media and authorities that categorize migrants in ethnic terms: Caucasian bandits, Chinese markets, Tajik construction workers, and so on (cf. Brednikova and Panchenkov 2001: 143; Varnavsky 2013: 469–470).

It is outdoor markets which have become signs of the entering of the ethnic factor into Russian cities. Basically, the evolution that the markets underwent from Soviet *barakholkas* and kolkhoz markets often took on ethnic forms, since it was connected with the influx of migrants and the import of foreign goods on an enormous scale. Labor migration and an influx of non-Russians constituted for many Russian regions a new phenomenon. During Soviet times, the amount of people from the Caucasus and Central Asia in e.g. Siberian cities was very small and there were virtually no Chinese people In the 1990s. however, the influx of newcomers became huge. All of the sudden, strangers who lived earlier just within the margins of the Soviet state started to move across its former protectorate, Russia. Many Siberian cities which, like Krasnoyarsk, had been closed during socialism due to strategic industries, were opened and moving to and out of them started to be allowed (cf. Dyatlov and Grigorichev 2013: 14–15).

Ethnicization is also about urban space. Markets not only became places where business was done, they turned into centers for minority life. The surroundings of the markets were adapted to their needs, with bars and restaurants for ethnic cuisine appearing in nearby houses, basements were adapted into warehouses, and so on. Foreign traders preferred to live close to their workplaces; they rented or bought apartments in the neighborhood of their markets, thus changing the composition of the local population. Open-air markets thus turned into important places of contact between newcomers and local inhabitants.

The reason why members of particular ethnic groups occupied whole branches of petty trade in post-Soviet Russia was that they were already there in the last years of the Soviet Union (cf. Il'in 1994: 195–196). The first trade minority one could notice to have appeared in the new Russia were the Caucasians,[46] who already during perestroika seized control over several branches of petty trade.

This was connected with the very negative stereotype of petty trade, and private economic activity in general, in Soviet society. Such a pejorative attitude towards

46 The term *Caucasians* (*kavkaztsy*) is very often used both in everyday language and the mass media. In the latter, mainly the quasi-juridical term *persons of Caucasian nationality* (*lica kavkazskoy nacional'nosti*) is used, usually with negative connotations. Thus, the term creates an impression that there is one, single Caucasian ethnicity instead of the myriad ethnicities that live in this region – or even that something like a Caucasian nationality exists. In the mass media, the term usually describes people from the Russian North Caucasus.
Dyatlov (1996a; 2000) describes two trade minorities in Irkutsk, namely the Chinese and the Caucasians, without separating the latter into particular ethnic or national groups. As he himself explains, it is justifiable in so far as his studies deal with public perception of the migrants. Since at the Krasnoyarsk markets there are very few people from the Russian North Caucasus, Armenia, or Georgia, I prefer to write about one particular national, or trade, minority: the Azerbaijanis.

trade and traders was, however, by no means a Soviet specificity. Throughout the history of human economic activity trading was carried out by strangers (Simmel 1971 [1908]). In a settled society, there is a need for a middleman who would provide the group with goods produced outside of it. Strangers to the group are not bound by kinship ties, status obligations, customary law, etc. Thus, they are allowed to engage in activities restricted for locals, such as trade, which is connected with mobility and transgressing various kinds of boundaries. These strangers may be despised, but at the same time they are needed. Because of this fact, "the stranger is still an organic member of the group" (Simmel 1971 [1908]: 149).

The idea proposed by Georg Simmel of what I would call a structural stranger can be applied to Caucasians in the late Soviet era. They did what the local population was unwilling to do and because of that they were both despised and needed. The economic experience gained in the late Soviet era enabled them to adapt quickly to the rapid changes of the 1990s. In this crucial time for development of market relationships in Russia, they managed to monopolize whole branches of trade and services. The same applies to the Chinese, who admittedly came to Russia later than the Caucasians but had the advantage over them of selling their own production.

❖ A conflict-prone trade

It should be noted that petty trade itself is conflict-prone. One of the reasons for this has to do with the personal nature of it. In a supermarket no one tries to haggle with a person behind the cash desk, because it is not this person who has set the prices. In petty trade, on the contrary, the conventional nature of the price is clearly visible. In addition, a customer may suspect that he or she has somehow been cheated by a trader when it comes to weighing, calculating the total amount to pay, and so on. Consumer goods, especially one's eating and one's money, are very sensitive issues. At Krasnoyarsk markets – and in Russia in general – matters such as the quality of goods, their source, and their real amount are highly uncertain for customers and become more a question of trust in a particular trader. Before such trust is established, however, a conflict can easily occur.

The very existence of trade minorities is conflict-prone as well. Minority vendors differ from the local population in race, physical appearance, and cultural habits. They have adapted to local society to a varying extent and their norms of behavior may violate local rules. As mentioned above, in Russia an additional factor is constituted by a very negative stereotype of private trade. A trader is perceived as someone preying on other people's work, someone who does not contribute to the public welfare. When Russian public opinion connected this depiction with the ethnic dimension of petty trade, the result was that *they* are making money on *us* (cf. Dyatlov 1996: 318; Dyatlov 2000: 33–34).

This is the feeling not only of customers but also of people who work for migrant employers. In Krasnoyarsk this was the case, first of all, of the Krastets. The owners of a significant part of the market were Chinese entrepreneurs, who usually employed Russian women as sellers. There were three main reasons for that. A foreigner can run a company in Russia, e.g., own a shop. However, the foreigner is not allowed to work in the shop if he or she does not have Russian citizenship or a residence permit (*propiska*). In addition, many customers preferred to buy from Russians; a Russian seller conveyed the impression that the stand is also owned by Russians. Last but not least, a customer would experience no problems in communication with the salesperson. Admittedly, the employers paid their employees, but the women had a feeling that the real profit from their work goes to foreigners.

In the 1990s, the presence of trade minorities became one of the symbols of *bespredel*. This was especially the case with Caucasians. They were clearly associated with the rise in criminality and violence. In polls Russians named as primary characteristics of Caucasians such features as loutishness, aggression, unwillingness to assimilate, and sexist attitudes to women (Dyatlov 2000: 145–147). Over the course of time these and other stereotypes and phobias were only confirmed by the wars in the Caucasus and various terrorist attacks. The fear associated with the Chinese, in turn, was attached not to their character but rather to their quantity. The "yellow peril" (*zhëltaya opasnost'*) supposedly looming over Russia, referred to the idea that the Chinese would colonize the country in a peaceful way by gradually taking over business domains, marrying Russian women, buying up land, and so on. Siberia with its vast and sparsely inhabited territories is said to be particularly prone to this danger (cf. Dyatlov and Grigorichev 2013: 29–30; Kovalskaya 2010; Blyakher and Pegin 2010).

As the fear of the "yellow peril" shows, Russians became worried about the growing influx of migrants. These migrants are very visible in urban landscapes and everyone knows there are many of them, yet their real amount is difficult to assess. This is because of the great scale of illegal immigration to Russia. In addition, numbers have a tendency to grow in press reports. The estimated amount of Azerbaijanis in Saint Petersburg in 2001, for instance, ranged from 50,000 up to half a million (Brednikova and Panchenkov 2001: 133). The quantity of Chinese official contract workers in Russia in 2008 was 331,000. In press reports, however, the number was assessed as high as even 8 million (Larin 2008: online).

To fight against illegal immigration, dragnets at open-air markets were frequently launched. These police inspections were intended to check everybody who worked there and – as it was in the case of Tanya – they usually served as a means to confiscate goods or extract bribes. Since 2002 various laws restricting the labor migration influx have consequently been put into effect. The most recent restrictions oriented on increasing state control over immigration processes were introduced in the beginning of 2014. According to estimates of the Federal Migration Service, in 2013, 11.3 million foreigners worked in Russia, of whom about 3.5 million did so illegally. The latter group comprised mainly people who had stayed in Russia longer than they were allowed to according to visa regimes. The aim of all this regulation was not only to reduce the influx of illicit migrants but also to make the legal ones settle in Russia (cf. Jarzyńska 2014: online).

Russia's economic difficulties of the 1990s and 2000s often caused labor migrants to be seen as the cause of all evil. Social tensions over living standards and competition at the markets frequently turned into ethnic conflicts. Again, one can say that the conflict was ethnicized over the course of time. The mass media and authorities on different levels also infamously contributed to the growth of the problem. Anti-migrant populist rhetoric is an easy way for a politician to gain popularity. In addition, sometimes social tensions are also used to further business interests. This happened during the state of emergency that followed street fights for the national parliament on October 3–4, 1993. Some of the decisions introduced at that time as law-enforcement measures were in practice the state's force being used to expel Caucasian traders from Moscow, because they posed competition to companies linked to the municipality. Since that time, Russian cities became a stage for frequent beatings, killings, and pogroms of labor migrants. It was usually a crime committed by a migrant that served as a spark that set off public demonstrations and riots against the newcomers (cf. Dyatlov 1996: 319; Dyatlov 2000; 7–8, 155–156; Jarzyńska 2013: online; Marciniak 2001: 526–527; Zarakhovich 2006: online). Also today, the massive inflow of migrants and the tensions it causes constitute one of Russia's main social problems.

Several mutual stereotypes between local population and migrants groups have come about. In colloquial Russian labor migrants from the Caucasus, Central Asia and China are contemptuously called *blacks* or *churki* (*wooden stubs*). The latter expression concerns their alleged stupidity. They are accused of committing crimes, dealing drugs, or littering the cities. The Russian population also complains about migrants taking jobs away from them and about foreigners spoiling the labor market by accepting substandard wages. The migrants, in their turn, see Russians as lazybones and drunkards. *They do not work themselves but they do not let us work*, one Azerbaijani trader told me, commenting on all the regulations that had been imposed on petty traders.

By no means, however, are the markets a source of only pejorative stereotypes. Positive experiences are also likely to appear on the basis on interethnic trade contacts. For instance, Andrey, a young Russian man, stated that he often prefers to buy from immigrant vendors. According to him, unlike Russians, they have a real flair for trade and thus the service they offer to customers is of high quality:

> *Because, the way I see it, citizens of Southern countries are better traders. Say you're strolling through the Krastets and you see two sellers, one Russian, one non-Russian. Well, it's simply more pleasant for a customer to buy from the non-Russian – and it's not just me, many people I know say that. He simply knows how to do it. It's a pleasure to buy from him, and this is where you'll go, not to a Russian* (V8).

❖ Wholesale trade in fruit and vegetables

At the time of my fieldwork, three neighboring markets – the Yuzhny (Southern), the Luch (Ray), and the Briansky – constituted the heart of Krasnoyarsk's wholesale trade in fruit and vegetables. The three markets were of crucial importance for the whole region. They served as a vital point of supply in fruit and vegetables not only for the city itself but also for the north of the Krasnoyarsk Krai and the southern republics, like Khakassia and Tuva. They were also located on a junction of a trade route from Central Russia to the Russian Far East. Besides their supply role, the markets were the region's main distribution center.

In general, people who worked there could be divided into big and petty wholesalers. A big wholesaler orders shipments from outside the region, whether from western Russia or from abroad. Some of them also offered goods from big farms they run in Krasnoyarsk Krai. They sold directly from trucks or from warehouses they rented at the market. Petty wholesalers, in turn, supplied themselves from the big ones. They worked in stands and resold the goods to retailers. It was only the latter group, the petty traders, who finally sold the goods to end customers in shops and stalls. If a big wholesaler ordered a shipment of 60 tons of apples from Moldova, a petty wholesaler bought from him a few hundred kilos and resold a few dozen kilos to each of several retail traders. This is how the system worked. A petty trader usually did not buy the goods directly from the big wholesaler. The latter would not be willing to sell in the small quantities the petty trader needs. Even if he did, the price would be higher than for a petty wholesaler. It is the amount of goods required that shapes the possibility of negotiating the price.

The main actors at the markets were Azerbaijanis, Tajiks, and the Chinese. Some representatives of other national groups, like Uzbeks or Armenians, also worked there. However, as if to confirm Nadya's statement quoted above, there were virtually no Russian traders. Along with ethnic differences some spatial and branch divisions could also be observed at the markets. Azerbaijanis specialized in *yevrofrukty*, literally "European fruit", i.e. fruit and vegetables imported from the West, such as bananas. This specialization was characteristic for Azerbaijanis across Russia in both wholesale and retail trade. They were even said to have a monopoly over it, though in practice the situation proved to be more complicated. Nonetheless, at the wholesale markets almost all Azerbaijanis, with very few exceptions, did sell *yevrofrukty*. Many of them – over a hundred, as I was told – came from a single town in Azerbaijan. The majority of the hundreds of Tajik traders, in turn, came from the region of Istaravshan. They sold mostly locally grown vegetables such as potatoes, onions, and beets, as well as fruit imported from Central Asia. Finally, the Chinese, who occupied a separate part of the Luch market, sold vegetables grown by them locally, mainly tomatoes and cucumbers.

There were some tendencies and even monopolies, but in fact who sold what was governed by a very dynamic and complex system. The supply of Siberia with foodstuffs depended on many factors: different sources of import, seasons, prices, and sometimes political decisions as well. In some spheres the three main groups competed with one another, but in general they seemed to have managed to work out an equilibrium. Tomatoes in big wholesale, for instance, were as a rule sold by Tajiks from May to July, by the Chinese from August to October and for the rest of the year by Azerbaijanis. It all depended on the season, and with it, it depended on the place of origin of the produce: the Tajiks imported the goods from Uzbekistan (from Tashkent, as the traders said), Azerbaijanis from south-western Russia and Turkey, and the Chinese grew them locally. Moreover, within wholesale, the groups left some space for one another. The Chinese, for example, did not interfere in the citrus market and Azerbaijanis did not sell local vegetables and dried fruit.

These were, however, very general and implicit rules and in practice individual traders competed with others within many branches. This was the case especially of the Tajiks and the Azerbaijanis, because some fruit, like grapes, can be imported either from Central Asia or from Turkey. On the other hand, many Tajiks came from the northern part of Tajikistan. Because of this, they often spoke Uzbek, which, in turn, let them communicate with Azerbaijanis, speakers of a related Turkic language. This made their inter-ethnic cooperation easier. There were also no obstacles for a Tajik, for instance, to order a shipment of apples from southern Russia. Space was left for individual initiative. The tendencies one could observe resulted more from the way in which individuals entered a business and the context thereof, i.e. personal ties and patterns of economic behavior.

❖ The embedded competition

Once, when I was working with Ibrahim (T17) and Kamran (T20) at their fruit and vegetables stand at the Slavyansky market, a Russian woman came and asked where the tomatoes came from. *They are local* (*mestnye*), answered Kamran. The woman insisted, however, *What does local mean? Where were they grown?* Kamran told her, *They are from Minusinsk*, with a tone of hesitation in his voice. The Minusinsky District in the southern part of Krasnoyarsk Krai – because of its specific warm microclimate – is a place where a lot of farming is done. *These tomatoes must have been grown by the Chinese*, the woman hissed and went away. Kamran's sad comment on this situation was, *They force us to lie*. If he had told the customer that the tomatoes were from Uzbekistan or from southwestern regions of Russia, he would have probably convinced her to buy them.

Both Kamran and the Russian woman were right – the tomatoes were local and at the same time they were Chinese. In recent years Chinese have leased a lot of land in the neighborhood of Krasnoyarsk. They have cultivated the land, often putting up greenhouses, and supplied the market with cheap vegetables. Views on their production were extremely divided. The traders, like Ibrahim and Kamran, said nobody but the Chinese – who were exceptionally hardworking – could grow such quantities of vegetables in a Siberian climate and deliver them for such reasonable prices. In public reception, however, vegetables grown by the Chinese were nothing but *sheer chemistry*. People said so much pesticide and artificial fertilizer was used on these vegetables that they were poisonous.

Cheap vegetables were very much desired in Siberia, however. The production of local people – based first of all on what they managed to grow in their gardens and dachas – was very low. Many foodstuffs needed therefore to be imported and because of transportation costs, they were much more expensive than in southwestern regions of Russia. The Chinese then met the local demand as they produced a lot and locally, and not only around Krasnoyarsk. As a consequence, customers at Siberian open-air markets – who for the most part came from the unprivileged classes of society – bought vegetables grown by the Chinese somewhat indirectly. It was petty traders who supplied themselves through the Chinese wholesalers and resold the goods at the markets. Thus, final customers usually acquired the produce unknowingly because traders typically did not reveal the real source of their goods.

As a matter of fact, however, there could be almost no other choice than to buy at least some of the vegetables from the Chinese. From August to October, when there was the season for Chinese-grown tomatoes and cucumbers, virtually nobody else traded in those vegetables in wholesale.[47] It was possible to order a shipment of tomatoes from another region of Russia, but these would be much more expensive. In September 2013 Ibrahim and Kamran sold tomatoes at their stall for 20–40 rubles per kilogram, depending on the sort and quality. At other stands the prices were similar. Tomatoes from Kazan, Kamran told me, would cost 65 rubles and it would be difficult to find customers willing to buy them.

---

47 Interestingly enough, I have never seen a Chinese trader who sold vegetables in retail. They did it only in wholesale. This was probably due to the yellow peril discourse and the very negative connotation which Chinese produce had in Russian society. Selling only in wholesale meant not being exposed to public view and could be interpreted as part of Chinese strategy: "the less contacts, the less conflicts" (cf. Dyatlov 2000: 156). To wholesale markets only petty vendors or owners of shops came. A wholesaler would not sell retail amounts of goods, so there was no point for an average customer to go there.

Admittedly, at the same time and at the same market pensioners sold tomatoes they grew in their gardens and dachas for even 150 rubles. Cucumbers, which the Azerbaijani brothers offered for 30–35 rubles, the pensioners sold for 100–120 rubles. In fact, these were not the same cucumbers. Pensioners sold their own produce which had cost them a lot of effort and which was very poor in terms of quantity. An average pensioner's tomato harvest was at most a few dozen kilograms. Such *real* (*nastoyashchy*) and *ecologically clean product* – as people described these vegetables – had to be expensive and some customers were ready to purchase them. The majority of customers, however, for everyday use preferred to buy tomatoes produced in local farms.

Brednikova and Panchenkov (2001; 2002) maintain that migrant petty traders work only according to purely economic principles and standard market competition. In polemics with other scholars (cf. Il'in 1994: 195–196; Snisarienko 1996: 146), they argue that Caucasians in Saint Petersburg – and ethnic minorities in general – have not monopolized any branches of petty trade. In their opinion, migrant traders are guided solely by their will to maximize profit in the shortest possible time. The relationships among members of one ethnic group are shaped solely by the economic rationality of individuals, by their calculation of gains and losses. The contacts between newcomers and local entrepreneurs are also of an ordinary business character: if they compete, they do so in a fair manner. Ethnic solidarity does not play any role, and thus one cannot speak about ethnic economy at all but only about market principles.

Following those latter authors, one could say the monopoly the Chinese had over a part of the market – tomatoes and cucumbers from August to October – was the result of purely market principles. It was their diligence, cheap labor force, and the fact they grew vegetables locally that allowed them to produce goods much more economically than anyone else. The monopoly was simply periodical and limited by a very complex system of deliveries to Siberia, so it did not disturb usual market principles.

The very fact of the existence of a monopoly, however, had consequences. A monopoly means that supply is necessarily limited – people who, for instance, would like to buy Russian instead of Chinese tomatoes do not have such an opportunity. In addition, in the Chinese case the system of pricing and competition seemed to be influenced by the ethnic factor as well. With Ibrahim and Kamran we went once to the Luch wholesale market's separate part occupied solely by the Chinese. We bought tomatoes for 22 rubles per kilogram. Kamran told me then:

> "*Even if an Azerbaijani managed to import cheap tomatoes from somewhere and started to sell them here for 20 rubles, all the Chinese at the market would suddenly start to sell their tomatoes for 15 rubles just to drive the stranger out of business*" (T20).

This meant that because of their strong group solidarity, the Chinese set fixed prices for their goods. According to Yakterina (T7), *we have diasporas emerging* [at the Krastets market], *that is – there's the head of the Chinese, the head of the Tajiks, the head of the Vietnamese.* The case of the Chinese at the Luch was probably similar: an informal leader, who supposedly commanded over mafia-like structure, set prices for all of his countrymen in trading. The two phenomena – limited supply possibilities and limited competition – made the tomato wholesale market in Krasnoyarsk strongly influenced by the ethnic factor.

Analogous situations took place in retail trade as well. This was the case of the Slavyansky market and the section described by traders as dried fruit (*sukhofrukty*). Besides goods such as raisins or dried apricots, the section also accommodates traders in various kinds of nuts and seeds. It is usually Tajiks who are engaged in this sphere, as many of these goods are imported from Tajikistan. At the Slavyansky there were three Tajiks who run stands with dried fruit. There was competition between the three men. Each of them had his own family to provide for, so they were keenly interested in the maximization of their profits. At the same time, the competition had its limits. Unlike the Chinese, who seemed to have some prices fixed for every member of their community, the Tajiks set their prices individually. They could do that, however, only to some extent. "*If I notice that one of them has lowered the price of a product too much, I go and I get the idea out his head*," Feyrudin (T38), one of the Tajiks, told me.

The other two Tajik traders were in a better situation than Feyrudin. Their relatives had farms in Tajikistan and sent them shipments, so they could sell goods directly from producers and at lower prices. Feyrudin, on the other hand, had to buy all the goods at Krasnoyarsk wholesale markets. Nevertheless, the other two cooperated with Feyrudin and did not dispute his requests concerning the prices. Thus, contrary to Brednikova and Panchenkov's statement, individual rationality was in this case limited by ethnic solidarity.

While lowering a price might have proven profitable for a single *sukhofrukty* trader, the whole group – including himself – would lose as a result. First off all, it is uncertain whether the price reduction would be beneficial. You could decide to sell one kilogram of dried plums with a profit margin of just 20 rubles instead of 50. Then, however, in order to make the same profit you would have to sell 5 kilograms instead of 2. It is not obvious at all that there will be demand for such quantities of dried plums. Only some of the customers walk around a whole market before shopping in order to check every price. Usually, they go directly to *their* sellers, to the people at whom, for one reason or another, they have chosen to shop. The fact that a few stands further they could buy cheaper plums – even if they have noticed it – might not be enough to stop them from doing their shopping at their favorite stall. For a trader it is not so easy to take over somebody else's regular customer. Last but not least, what is undesirable for all traders of the same branch is a price war. If you decrease the price of a product, then your neighbor can cut the prices of two of them, so you are forced to make further reductions. Such competition can be enfeebling for everybody. Thus, it is better to hold reasonable prices that would suit both traders and customers.

In addition, the Tajik vendors in dried fruit had one more goal in common – they did not want anybody else to open a stand with dried fruit at the Slavyansky. "*If somebody came, we would make him give up*," said Feyrudin. He meant that, together with his countrymen, they would use more or less violent means in order to make a stranger leave the market. Competitive price setting – even dumping– would be the first step. He hinted that such cases had happened before.

In general, Tajikistani traders in Krasnoyarsk showed many signs of ethnic solidarity. There were some 10 Tajik entrepreneurs at the Slavyansky and they maintained close ties with each other. Even those who worked within the same branch cooperated, as the practice of price-fixing demonstrated. Once, when I came to talk with Feyrudin, he was keeping an eye on the stall of his neighbor, a Tajikistani man who had gone to pray at a mosque. Feyrudin served the neighbor's customers and left all the earned money for the friend. Besides this, he often supported his countrymen. This applied to his fellow tradesmen, but he and other Tajiks also helped strangers. From time to time, for instance, there were Tajik people coming to the Slavyansky looking for financial support, whether to go back to their homeland or instead to set themselves up in the new place. All the Tajik traders would then chip in, contributing 100 rubles each. "*If somebody is reasonable, then he finds a job for himself. If he is stupid and drinks his way through the money, then we will not help him anymore*," Feyrudin told me. He also frequently helped his relatives or friends from Tajikistan when they came to Russia to work. He himself did not have the possibility to employ anybody, but he took such a person to his home, gave them tips, and lent him or her some money. "*Usually three days are enough to get one's bearings. People then come to give money back and to express their gratitude*," he said.

Obviously enough, migrant businesses were indeed businesses in that they were run according to the main goal of every entrepreneur: making a profit. Furthermore, general laws of supply and demand undoubtedly worked at Krasnoyarsk open-air markets, while competition between various minorities usually operated according to standard market rules. The question remains, however, what the general market situation created by the existence of trade minorities is. More specifically, one can ask what the competition looked like within particular ethnic groups. Here one can find spheres in which profit-oriented economic behavior was considerably imitated. Thus, the ethnic – or more generally, the communal – factor mattered in the way migrant traders entered markets and built relationships with other agents there.

The way in which migrants took up their activities was ethnicized. For instance, the system of hiring at the wholesale market ran solely according to ethnic ties: Tajik entrepreneurs employed only Tajik people for carrying or sorting their goods, and so on. In addition, the situation of migration brings with itself an actualization of ethnicity: suddenly one becomes a stranger. This takes place on a personal, subjective level – how a person feels and identifies himself or herself – but at the same time also on an objective level – how the person is perceived by the local population. Through this objectivization, personal ties gain an ethnic character. This, in turn, leads to identification with an ethnic group, which in itself can already have economic consequences. Once you have entered a business through informal (family or friendship) channels, you are supposed to obey the standards of the group you have joined. Your enterprise constitutes a part of a bigger entity, whether a branch of trade or all your countrymen trading at a market. This influences the way you run your business; for instance, you are not supposed to lower your prices too much or to intercept other people's customers, unless you want to be despised by other members of your minority. The group obliges you also to support its members and this sometimes affects your choices of supply sources. As a result, your desire to maximize profit is regulated by different obligations.

If we fought against one another, we would not survive I was told by Lyuba (T59), the trader described in Chapter 4, and her friend Valentina (T62). The two women worked just opposite each other at the Slavyansky and traded in similar foodstuffs. However, the first specialized more in macaroni, while the second had a rich variety of rice and groats. If they really competed with each other, probably one of them would win and the other lose. This was, however, undesirable since they had known each other for a long time and had become friends. Therefore, they decided to specialize in narrow spheres. If one of them did not have a certain good a customer asked for, she would direct the buyer to her friend. For them, it was not maximization of individual profit but rather common survival at the endangered market that became their main goal.

In contrast, the competition between Fedya and his Russian neighbors – Sergey and Lena – led to the failure of the former. As a matter of fact, the competition might have been much softer. Sergey and Lena occupied a bit different niche from that of Fedya and other fruit tradespeople, as they sold premium quality goods at high prices. Their target group were well-off clients who did not want to buy from migrant traders. Other fruit traders – all of them migrants – targeted rather the middle or unprivileged social classes. Nevertheless, Fedya had very strong opinions on Russians. Sergey and Lena, in turn, were very critical about the inflow of migrants to Russia. Their relationships with other migrant traders were rather cool, not as hostile as with Fedya. It seems that several factors – including matters of personal characters – turned the competition between them and Fedya into a conflict on an ethnic basis.

The usage of magical means by both sides of the conflict can be viewed as helplessness in the face of the overall situation. The competition in the branch of fruit and vegetables at the Slavyansky was tough and the number of customers coming to the market was decreasing. Having exhausted standard ways of developing their businesses, both sides started to seek support from spiritual powers. For Fedya turning to a *babka* and the use of magic – which strongly influenced his economic behavior – was oriented not at maximization of profits, since he constantly suffered losses, but at plain survival. It was the behavior of a person who was utterly cornered.

Contrary to Fedya, Sulayman (T30) – who has been already mentioned in this chapter – on the basis of his religious convictions, questioned the very idea of competition. Sulayman was called at the Slavyansky the "Mullah". He did not in fact belong to Muslim clergy but he was an ardent believer of Islam, had a long beard, and dressed according to religious code. He might have been viewed as a rival to Fedya, Sergey, and Lena because he sold fruit and vegetables just opposite them. He himself, however, did not perceive the situation in such way at all. He did not engage in any conflict and just counted on God's support. According to him, it was Allah who decided what happened to every apple or potato. It was Allah who had created the goods, so it was him who designed their end, whether they would decay, be eaten, or give a beginning to new plants. If a customer bought something from Sulayman, it was thanks to God. If the customer did not, it was thanks to God, as well. From such perspective, the phenomenon of competition does not make sense – whether a trader is successful or not depends not on his or her personal effort but on God's will. In Sulayman's opinion, the only thing a trader must do is to be honest. You shall not cheat on balance or write that your tomatoes are from Tashkent if you bought them from the Chinese. Otherwise, God would remove his grace and then you would fail for sure. Such was Sulayman's position.

Building communities at Russian open-air markets

In practice, every entrepreneur chooses the strategy he finds best for himself, so the economic strategies of tradespeople form a wide range of possibilities. Nevertheless, some tendencies can be drawn here. The longer one works at a particular open-air market, the more likely he or she is to use softer forms of competition in order to establish better relations with other actors. To build a community seems often to be a more far-sighted economic strategy than to seek just one's own interest. If you develop a peaceful relationship with your competitor, both of you can profit from common pricing, for instance. You are also stronger in coping with common problems. After all, it does not make sense to win against your rival, if on the very next day you will lose as well. It is better to survive together.

In the case of the Slavyansky the goal of collective survival became particularly urgent when the market was reportedly doomed. In addition to many problems the tradespeople had already faced – difficult work conditions, the competition posed by supermarket networks, and the financial crisis – the risk of losing their common workplace appeared. All these factors contributed to their sense of general uncertainty about the future. It was in Tanya's community, a community that consisted mainly of Russians who had worked at the market for several years, where a defense initiative was started. Signatures were collected and a few petitions were brought to the district administration. This action, undertaken in common interest, united different groups at the market. It is the ability to organize for the sake of common goals that creates a feeling of belonging together, mobilizes social capital, and constitutes perhaps the best proof of a community's strength (Djelic and Quack 2010: 5–7; Granovetter 1973: 1373; Lin 2001: 12).

However, the use of the notion *community* to describe tradespeople working at an outdoor market carries with it some risk. The term, unless specifically defined otherwise, presupposes close-knit relationships, rich emotional ties, and physical proximity between members of a group (Djelic and Quack 2010: 3). Such requirements are not fully met by people who work at an open-air market. Their relationships are seldom close-knit and emotionally rich. Moreover, their participants are separated by different ethnic backgrounds and often contradictory economic interests, as they pose a competition to one another.

On the other hand, however, what unites tradespeople at an open-air market is their permanent coexistence. Their communality also goes far beyond that. In the case of Russia, these people have a common feeling of sharing similar life trajectories and tackling similar everyday problems. The unwilling entrepreneurs found themselves trading at a market as a consequence of perestroika and because they were forced to do so by vital necessity. They are not simply people who happened to earn their living in the same way. They belonged to social layers which, during the systemic transformation, had been excluded from the main course of the country's development. This very fact forms a common experience and allows for the appearance of a sense of community.

Surprisingly enough, this sense applies also to labor migrants. Admittedly, newcomers form a serious competition for local inhabitants not only at outdoor markets but also in other sectors of the labor market. Nevertheless, as I repeatedly heard from Russians, *migrants did not come here from a good life*. Migrants were forced to come to Russia and to take up – just like local people – work at an open-air market in order to sustain their families. "*After all, they have to feed their children, too,*" I was told by Lila, a Russian tradeswoman (T51). Thus, there is a sense of compassion, even though migrants are usually accused of taking work from Russians and spoiling the labor market by accepting substandard wages.

The Slavyansky allows one to observe the different manners in which various communities are bound together. An instance of this are Tanya (T4) and Alik (T12) – the man described in Chapter 4 – who belong to two different types of community. In the case of Alik, his community was based on ethnic ties and strong group solidarity. Practically all Alik's friends and acquaintances were Azerbaijani. He started cooperating at the market with his Azerbaijani friend. A driver he frequently cooperated with was Azerbaijani as well. Last but not least, Alik often used to support other Azerbaijani tradespeople, mainly by lending them money. He also invited them to cafes and organized parties. All this allowed him to become a kind of a group leader. That is why he had no problems finding people willing to help him when he had been cheated. What is to be noted here is that the man who breached the contract was a Russian, not an Azerbaijani, whose obligations to keep the contract would be much stronger.

Tanya's pattern of behavior was different. Of course, she did not have enough power to kidnap the employer who did not pay her salary and extort the money from him. However, she successfully took up softer measures, as the example of finding an informal source of acquiring newspapers has shown. What is more, Tanya managed to establish good relationships also outside her ethnic group, thus building a community on a transnational level. She became a kind of bridge that connected different communities. In opposition to Alik, who had worked at the Slavyansky market for just a short time, the relationships established by Tanya were based primarily not on ethnic ties but on long-lasting communality and mutual aid. This became clear when she was in need and Nurdzhigit gave her a hand. She, in turn, could prove her reciprocity by selling his goods when he left the market.

Alik lived in a small and relatively closed ethnic community that was bound by strong ties. After his unfortunate story with importing wood to Azerbaijan, he preferred to make business relationships that would be based on stronger obligations than just economic interest. It was his cousin from China who became his main business partner. As he had worked at the Slavyansky market for a short time only, he did not go outside the community of Azerbaijani traders. Tanya's community, in turn, was bigger, transnational, and bound by rather weak ties. Members of this community had known each other for years, at least by sight, but usually would not spend their free time together. The simple principle is that with a growing number of members of a group, its social capital also increases, because this capital is constituted by resources embedded in social networks, norms, and trustful relationships (Lin 2001: 3ff.; Putnam 1993: 167). This was the case of Tanya's community and it became visible when a tradeswoman whose husband had died received financial support even from people who did not know her personally. Paradoxically enough, it is the weak ties that build bigger and overall stronger communities. The strong ties, on the contrary, "breeding local cohesion, lead to overall fragmentation" (Granovetter 1973: 1378).

6. Conclusions: an insight into the political economy of Russia

Political and economic processes in 1990s Russia have been characterized not as a transition from socialism to capitalism but from socialism to neo-feudalism, especially at regional levels. Scholars demonstrated what appeared to be feudal suzerainties of a sort within the state economy and the privatization of its property, state enterprises, public administration, and mafia structures. In a critically unstable political and economic situation, individuals had to rely on personal connections rather than institutions, which made them dependent on local patrimonial systems (Burawoy 1999; Burawoy 2001: 270; Clarke 1992; Humphrey 1991; Humphrey 2002b: 5–20; Verdery 1996: 204–228).

Nevertheless, neo-feudal suzerainties were something characteristic more of the first years of the transition and more of Russia's hinterlands, because its centers assimilated rapidly with the global economy (cf. Burawoy 2001: 287). In general, Russia was treated – as the term *shock therapy* mutely implied a country and its inhabitants should be treated – with neoliberal capitalism in its sharpest form. While the metaphor of neo-feudalism captures internal relations of power, it fails to grasp the ways in which Russia and other postsocialist states were colonized by international capital. When the market was opened in Russia, many corresponding institutions that could somehow regulate it – such as a banking system, stock exchange or arbitrary courts – were nonexistent. The therapy was also not preceded by any attempts at building a civil society in the country. The recipe was to let the *invisible hand* operate freely and that turned out to be disastrous for masses of people. The capitalism which appeared as a consequence was defined as a *bandit, oligarch, nomenklatura,* or *wild* one (Afanasyev 2001: 166–168, Nazpary 2002: 9). What is more, President Putin himself is said to be an ardent follower of such a model, one in which capital would not encounter any restraints, in which oligarchs would be bigger capitalists than the Western ones and, finally, in which there would be a "maximum oneness of state and business" (Pavlovsky 2014: 57).

One of the most striking effects of the postsocialist implementation of capitalism was a widening of social disparities. It stood in stark contrast to socialist times, when social gaps were relatively small. Systemic transformations in the Eastern Bloc led to widespread pauperization on the one hand and to the creation of narrow nouveau-riche elites on the other. In other words, two new and extremely polarized social classes were created. In general, one cannot speak about the existence of any significant middle classes in postsocialist societies. As a matter of fact, they appeared practically only at later stages of the changes (cf. Mokrzycki 1996), when national economies bounced off the ground after their initial postsocialist collapse. In Russia, even though the underprivileged, the dispossessed, or the losers constituted in fact a very heterogeneous class, in their own eyes the distinction between *us* (*my*), *the poor* (*bednye*), *the people* (*narod*) in opposition to *them* (*oni*), *the rich* (*bogatye*), *the new Russians* (*novye russkie*), and *the authorities* (*nachal'stvo*) became the main tool of social classification (Nazpary 2002: 16).

What happened in Russia was "the transition from a nomenklatura to a merchant financial bourgeoisie, centered on banks, trading, mafia, and exploitation of natural resources" (Burawoy 2001: 286). It was the class of the New Russians, i.e. of market intermediaries, which became the main beneficiaries of the transformation. It comprised people who did not generate any new resources but only lived "off the rapidly diminishing and impoverished productive classes" (Burawoy 1999: 14). In the scenario of development that 1990s Russia followed, liberalization in practice meant the disorganization of state structures and lack of any control over socioeconomic processes. The top tier of the New Russians, people who came to be known as the oligarchs, seized the most profitable branches of the economy and used their influence to affect the weak state institutions, thus expanding their power also over politics.

One could notice here that petty traders similarly did not generate any resources. However, it is exactly them who constituted the diminishing productive classes. Although during the 1990s shuttle traders played an important role in supplying Russia with goods, they could hardly be called merchants who held control over the country's imports and exports. Moreover, taking individual entrepreneurs as a whole, one will find also traders who sell their own produce or people working in the service sector. The political economy of 1990s Russia remained unchanged in many spheres to this day. The fact that state structures have strengthened considerably does not alter the general picture: it is still the possessive classes who create policies oriented at securing their broad privileges. The sector of small and micro companies – despite a few national programs intended to develop it – receives only nominal support. Laws that concern it have been occasionally discussed over the years, only to end up contradicting other regulations and not forming any general vision for how this sector should operate (cf. Radaev 2011b; *Zakon o razvitii malogo i srednego predprinimatel'stva* 2007).

Meanwhile, it is big economic actors that take hold of an increasing number of segments of the Russian market. In the case of Russian retail trade, these actors are chain stores, big suppliers, international companies, importers, developers, real estate owners, and all those wealthy enough to invest in and run shopping centers. It is them who benefited the most from the act *On retail markets...* As various analyses have shown, various networks – very often international ones – are developing rapidly in today's Russia, in some cities monopolizing entire branches of retail trade (Radaev 2003; Radaev 2005, Radaev 2011a). With the serious limitations on the functioning of open-air markets, they achieved a victory over their competition, and managed to do so despite great social opposition. Thus, one can say that just like during the Russian systemic transformation, masses of people became dispossessed, turned into underprivileged social classes, and became tradespeople. Today they lose once again.

❖ The contribution of traders to Russian systemic transformation

In this study I analyzed the development of Russian petty trade since the time of perestroika until today, that is, until the end of 2014. In order to understand the Russian systemic transformation anew, I adopted a grassroots perspective of the petty trader. I claimed that it was also due to individual entrepreneurs and people involved in the informal economy under socialism – including traders – that the transformation was started. By *transformation* I understand the change in people's ways of behavior, their position towards the social change. In this sense – in the case of the USSR and post-Soviet Russia – *transformation* preceded *transition*, that is, the change of political and economic institutions. After all, it is actual people who create and steer institutions. Obviously enough, it is also actual people who often have to adapt themselves to systemic changes imposed on them above. This was the case of petty tradespeople in postsocialist Russia, too. At the same time, however, the adaptation they performed was active and creative. In their masses petty traders played a great role in shaping the course of Russian systemic transformation.

One of the questions I asked in this context was about the sources of entrepreneurship in Russia. I see them in various informal economic practices that lay at the core of the socialist economy. Moreover, I tried to show that conditions of living in the everyday of real socialism forced an individual to develop an entrepreneurial position. One had to undertake risk and to be innovative, obstinate, and self-reliant. In order to solve their life problems, individuals had to *organize* things, to manage matters which often required usage of *connections* and *acquaintances*. These factors, along with the existence of spheres where private labor activity was allowed or indeed necessary, created a fertile ground for developing forms of entrepreneurship introduced in the USSR in the late 1980s. That is why when socialism fell, many former Soviet citizens found themselves ready to face the challenges of a completely changed reality.

One could call petty trade during socialism an island of free market relationships in an obviously non-market system, as this trade was organized based on the principles of supply and demand, price efficiency, and competition (cf. Sik and Wallace 1999: 699). However, the possibilities to be active were very limited and required undertaking illicit measures. One could also claim that the privatization of state and Party property carried out by communist officials or the functioning of the administrative market were likewise the beginning of a free market. Again, the informal and illegal character of these practices is obvious. As long as informal markets and economic relationships were squeezed into very specific conditions they could hardly be called a free market. Rather, they formed quasi-markets and quasi-entrepreneurship. Of course, my usage of this "half-" and "quasi-" language only goes to show how difficult it is to find proper notions to describe such a fluid and mysterious reality. The informality connected with private entrepreneurship that was inherited from the Soviet times pressed heavily on traders in the 1990s.

The decade of the 1990s can be called for masses of Russian citizens a tough school of entrepreneurship. As I was told by Nadya, with whose words this work opened and who had been a *chelnok*:

> *It's like they say: "If you want to teach someone to swim, don't come in and save them". And likewise here – if you want to teach someone to earn money, you shouldn't feed them. Make them go around hungry, and then perhaps they'll achieve something. And that's exactly what happened to us* (T5).

People managed to survive those difficult times thanks to their energy and resourcefulness.

Large-scale geopolitical events often make life for entire social classes exceedingly difficult. In this case, however, tradespeople – even the pettiest one who sold things out of their hands at the corner – managed not only to secure the existence of themselves and their families but also to work out effects measurable in macroeconomic calculations. In the 1990s Russians massively engaged in trade because of shortages, unemployment, lack of cash money, and so on. At the same time it was exactly the traders who neutralized these problems. They brought the products in demand, they created jobs for themselves, and they invested money at home. In this way, their adaptation to the rapid social and economic changes brought by the fall of socialism was by no means passive. On the contrary, they became active agents, who significantly contributed to overcoming the biggest problems of Russian systemic transformation.

❖ Informality and coping with uncertainty

According to the second claim I posed in this book, in the course of the systemic transformation, tradespeople – as well as the underprivileged classes of Russia in general – had to cope with uncertainty. This uncertainty stemmed not only from the usual conditions of trade activity, such as risk of loss, uncertainty of gains, and the difficulty of assessing future turnover and of making investments. First of all, it resulted from the civilizational catastrophe that 1990s presented in Russia and from the instability of the country's political and economic situation. For tradespeople, various uncertainties made up a general uncertainty of life, that is, the uncertainty of sources for making a living for their families and thus for the quality of these families' future existence. Engaging in trade and different strategies of running a trading activity can be viewed in this context as an effort to provide basic certainty.

Establishing some certainty for life proved in the newborn Russia to be difficult. This was so, among other reasons, because of the range of the sphere of informality tied to private entrepreneurship. Informality, I argue, resembles a stick: it has two ends. On the one hand, it can give an individual a chance to achieve success. Individuals are not exposed to legal restrictions so much, so they can fully use their skills, act through informal channels, and thus possibly make good money. On the other hand, true freedom does not exist and an individual always meets restraints, legal or otherwise. Being unrestrained by official institutions – legislation, law-enforcement bodies, public organizations, and so on – means also not being secured by them. Thus, in such a situation an individual is more prone to risks and uncertainties of an informal character.

For instance, Russia's state structures in the 1990s were weak. Both central and regional authorities found themselves incapable of implementing their policies and governing over the country's vast territory. As a consequence, Russia was dissolving into a multiplicity of privatized and independent domains of political, economic, and geographical character. What one can call the *weak state syndrome* of post-Soviet Russia resulted also from the fact that the state "rather than leading a social and economic transformation, was itself transformed by latent socioeconomic forces" (Stoner-Weiss 2006: 3). As such, it could not defend its ordinary citizens against destructive market forces. Since the state did not regulate the market – for instance, it did not enforce contracts or protect property rights – citizens had to withdraw from state structures and to try to influence the market in informal channels.

This proved to be a burden rather than a chance to achieve success, for the informality was often linked with criminal factors. If this crime was organized, the average individual found himself or herself particularly vulnerable. However, the burden of informality was not just about criminal aberrations. In a state which does not guard even most basic rules – such as securing salaries in the public sector, pensions or deposits in its banks – citizens face a general uncertainty of existence. For Russian petty traders it resulted in an anxious situation in which they did not know what could happen to their businesses or, for that matter, to them and to their households.

Two things are crucially needed to run a shop or another business: safety and predictability. In other words, you need certainty. You have to be sure, for instance, that your shop will not be robbed and if you order a delivery and pay for it that it will come at all and it will come on time. The Russian state in the 1990s was unable to fulfill these basic conditions. That is why what might be observed was substituting of the state by various non-state enforcers of rules. One could also say that state functions were diffused to other subjects. This included taxation, because entrepreneurs often had the choice of whether to pay taxes or remain in the shadows and to pay the mafia or some officials (cf. Humphrey 2002b: 100–101). As a matter of fact formality and informality overlapped to such extent that they could not be separated. Policemen on duty extorted money from citizens. Bandits, in turn, were elected to state authorities.

In a state suffering catastrophe, it is the rules that become essential; what bodies secure these rules is of less importance. The basic means to ensure certainty is money, whether from extortion or a bribe. "A good *krysha* means good business," says the gangster quoted by Ries (2002: 279). That means entrepreneurs, in order to survive, have to establish good or even trustful relationships with officials and racketeers. It is necessary to have somebody you can count on. According to Igor Klyamkin and Lev Timofeyev, corruption in Russia functions for an average citizen only as an economic phenomenon. For a very long time officials had to be *fed* and the administrative decisions he could take were goods to buy. Bribing then is in practice legally and morally neutral and signalizes the prevalence of customary law over state law. The habit of bribing officials or paying a medical doctor on the side for treatment was easily inherited from Soviet times. It is usual even for young generations that did not live under socialism. Indeed, entrepreneurs or people who want to receive good treatment do not have any choice but to bribe, they are compelled to do that, but they accept it since corruption supplies them with the desired services. After all, corruption is convenient and the costs of bribes can be simply included into the economic calculation (cf. Klyamkin, Timofeyev 2000: 90–98, 217–222)

❖ Embedded economy of petty traders

In the history of the Slavyansky market one can find a reflection of the history of petty trade in Russia in general. The market, which was created in a spontaneous way during the heyday of petty trade in the mid-1990s, quickly turned into an important commerce center for the local urban population. For those who had lost jobs in state enterprises and willingly or not became entrepreneurs, the market offered the possibility to make a living. "*Trade let us survive*", "*The market feeds everybody,*" I was told by Lyuba (59). In course of time the Slavyansky underwent ethnicization. With Russians gradual driven out of the market, different branches were seized by minorities – fruit and vegetables by Azerbaijanis and Tajiks, clothes by Kyrgyz, while groceries remained mainly in the hands of Russians. The Slavyansky prospered until after the turn of the millennium, when legislative regulations, the growth of supermarket chains, and the financial crisis started to leave their imprint. Combined, those negative factors brought disaster on the market.

From 2007 to 2013 the Slavyansky traders experienced what I called legal uncertainty, the situation in which the legal status was undefined. As far as the market constituted the main source of income for the vendors, its status translated into an uncertain future for their households. Tradespeople tried to cope with this uncertainty with whatever measures they had at their disposal. Uncertainty of life is something characteristic of contemporary societies, which are constantly shaped and reshaped, but it is particularly present in postsocialist countries, where the pace of social changes is especially rapid. Establishing relationships based on trust is something that brings about a general sense of safety, because they mean that one can expect to be helped in case of need. This happened to Tanya (T4) when she had to hide from the police or to Fedya (T9) when he needed to borrow money to pay the rent. Trust and community-building reduce the harmful influence of risk and uncertainty, thus creating opportunities for one's wellbeing in the future (cf. Putnam et al. 1993: 170; Sztompka 2003: 16–21).

Community-building has an impact on economic mechanisms. Monopolies of particular ethnic groups on particular trade branches, systems of price fixing, and limited competition – these phenomena show that at the markets of Krasnoyarsk, economy was embedded. Monopolies not only mean that power of supply is limited and thus the prices are not set by the *invisible hand of the market*, but they also result in restricted opportunities to enter a branch of trade. For some reason there are virtually no Russians in the wholesale trade in fruit and vegetables in Krasnoyarsk. Apparently, if you are a Russian entrepreneur willing to enter the market, you may encounter grave difficulties while doing so, and not only because of the usual market competition. The phenomena of price fixing and limited competition in turn need not be of an ethnic character. Such cases may happen between members of different ethnic groups, too, but they are more likely to appear within an ethnic minority. The ideal free market does not exist for many reasons and ethnicity is one of them.

The massive influx of immigrants in the 1990s and 2000s initiated new processes in post-Soviet Russian society. One can talk about an actualization of ethnicity for both locals and newcomers. The process of ethnicization was especially visible at urban open-air markets which turned into important places for intercultural exchange. It was at the markets, for instance, that mutual stereotypes and phobias were born. Nevertheless, by no means are markets only sites of ethnic conflict. In their everyday life they rather constitute places of peaceful coexistence and collaboration of people of different nationalities and cultures. Despite all the differences, the local Russian population and the migrants from the former Soviet republics share a lot in common. An ethnic-based community is the easiest to build, especially for migrants, who often do not reach beyond personal networks and remain within their closed groups. However, if people work together for a longer time, interethnic ties and multinational communities can also be established. Thus, markets can be places where different stereotypes are not only created but also challenged.

Consequently, tensions between the local population and newcomers are toned down by a kind of compassion. As many Russians admit, *migrants do not come here from a good life*. There is a common feeling of sharing the same life trajectories. As I was told by a Russian tradeswoman Lila (T51), *the story of everyone who works at the market is the same: perestroika came*. She meant by this not only Russian but also immigrant vendors. Perestroika constituted a uniting factor, for it determined the lives of both groups. If one connects these two quotations, the result will be that if everyone suffers because of perestroika and the dissolution of the USSR, migrants suffer more, because they are forced to leave their countries to work abroad.

At the end, one should attempt to make some predictions for the future and to point out some directions for further research. Today, as a consequence of the annexation of the Crimean Peninsula and the armed conflict in the Eastern Ukraine, Russia faces new economic instability. The situation is beginning to resemble economic competition from the period of the Cold War. The falling prices of oil in international markets, international sanctions imposed on Russia, and the import bans introduced by Russia in reply have brought a serious blow to Russia's economy. As a result, the inflation rate has substantially increased and the ruble dramatically fell in value. Undoubtedly, just like during the Global Financial Crisis, it will be the underprivileged classes who will suffer the most because of the country's worse performance. It is food prices that are increasing the most, and this will hit each and every household – the poorer the household, the more it will suffer.

The question then is as following: how will Russian citizens cope with these problems? If the economic decline continues, one can expect broad masses of society to be pushed down into the sphere of subsistence economy. This, in turn, will lead into an expansion of informal economic practices. It seems very likely that strategies of survival will be tied to petty trade yet again. What forms will the trade take up? How will the public's attitude to labor immigrants evolve? Will the vulnerable social groups be able to influence national policies? The new crisis might bring a loosening of regulations on individual entrepreneurship, and this in turn could herald a change in the political economy in Russia.

Appendix 1. Methodology

My research questions concerned ways in which an individual copes with radical social change. I wanted to know what survival strategies, especially those linked with trade, people adopted in the 1990s and what strategies they adopt today. My intention was to place the phenomenon of post-Soviet petty trade in its wider context: the economic attitudes and patterns derived from the Soviet era and the social perceptions of trade. I also asked myself what kind of relationships tradespeople at an outdoor market build among themselves: were these relationships only of a competitive character or did they also take on forms of cooperation? Finally, the aim of my project was to look at how Russia's dispossessed people, who remain on the margins of the state's policies, see themselves as the citizens of this state.

In order to find answers to these questions, I conducted ethnographic fieldwork in Russia. This, however, was preceded by theoretical preparation. As Burawoy rightfully argued, an ethnographer comes in close contact with the world of his or her study but does not perceive the world as it is. Contrary to the claims of naïve empiricism, there are no pure facts to which ethnography would grant special access. An ethnographer starts a study not with data but with a theory, a social theory to be precise. It is the theory which leads the ethnographer to the data and later, it is the data which leads the ethnographer to revise the theory that he or she began with. Theory is something both natural and necessary: "We cannot see social reality without theory, just as we cannot see the physical world without our eyes. Everyone carries and uses social theory, cognitive maps of the world we inhabit, although not everyone is a social theorist, that is, someone who specializes in the production of such maps" (Burawoy 2009: xiii). Theory then consists of understanding how the world works and the special task of a social scientist is to be particularly aware of how various theories are constructed.

Here one can find also an answer to the fundamental anthropological dilemma which Clifford Geertz put in a form of a succinct question: "Who are we to speak for them?" (2000: 95). If an ethnographer, an anthropologist, or a sociologist specializes in the production of cognitive maps, he or she should develop an analytical theory that would encompass the existing casual or commonsense theories. It is this special theory which distinguishes a researcher from the people he or she studies. Maps are of varying quality, they differ in scale and precision, and they can reflect the world in a better or worse way. Indeed, postmodernist critique effectively challenged the authority of the classical anthropologist, an allegedly transparent and undetermined specialist who was said to deliver objective representations of so-called primitive societies (cf. Clifford and Marcus 1986). This is not to say, however, that anthropologists should not master their field and cannot become real experts in the interpretation of cultures.

My trip started in September 2011 with a fellowship at the Moscow Higher School of Economics (HSE). I stayed in Moscow until January 2012. During this time, I attended seminars at the Laboratory for Studies in Economic Sociology at the HSE, which gave me a chance to meet and learn from the scholars who work there. Apart from this, I continued preparatory research that consisted in reading Russian literature of the subject, which is often unknown and unavailable in the West. I also made observations in the city's streets and conducted a few interviews with tradespeople.

Both the consultations with Russian scholars from Moscow – and later, in course of my trips, from Krasnoyarsk, Irkutsk, Novosibirsk and Barnaul – as well as reading the Russian literature on the subject were of utmost importance to me. This was not only due to the useful information I received but also because of a methodological assumption I had adopted. Namely, in creating my analyses, I tried to maintain a balance in drawing on both Western and Russian authors in order to balance out the opinions and conclusions both sides tend to arrive at. I found this important insofar as Western-originated studies on transformations in postsocialist countries carry a risk of strengthening Occidentalism and reproducing hegemonic political discourses. Many such works have been produced without knowledge of any local language, relying on sources written exclusively in English. Even if native scholars were included in bibliographical references, quite rarely were local ethnographers and their fieldwork mentioned. Moreover, the quantity of external analyses of postsocialist transformations is not necessary matched by their quality (cf. Buchowski 2004).

In this context one might ask a general question: to what extent do theories and notions worked out by Western social sciences apply to contemporary Russia, a Russia which is neither the West nor the East but simply itself? Afanasyev named two basic factors which as a rule are ignored by those who try to grasp the phenomenon of today's Russia. The first is constituted by the fact that there are two opposite yet complementary realities which form the content of human life in Russia: the *normative* reality and the *actual, gray,* or *shadowed* (*tenevaya*) one. The second factor is comprised by "the presence in Russian reality of phenomena, institutions, offices that are unnamed or whose names do not actually describe them (...) There are phenomena or institutions which nominally have long existed in our country – say, the state, parliament, prosecutor's office, politics, courts of law, economy, business, civil society, the market and so on. But in the meantime, temporarily, before being properly *christened*, they were given names that belonged to other things. The names which in other places and other times denote not entirely the same, or indeed entirely different, phenomena" (Afanasyev 2001: 16).

As Chris Hann put it, "the insights of close-up fieldwork observation are especially valuable in periods of uncertainty and institutional instability" (2002a: 7). Thus, I also argue that ethnography is especially useful in attempting to describe such dynamic phenomena as petty trade in contemporary Russia. This is precisely because such trade was done to a considerable extent within the gray or informal economy, which is a part of a bigger entirety – the gray reality. Both the gray economy and the gray reality fail to be described by official statistics and are regulated by state legislation only to a certain degree. Quantitative research cannot grasp them but qualitative (e.g. ethnographic) research is able to do so. Moreover, the anthropological methodology is designed to ensure that the image of petty trade thus produced is grounded in fieldwork experiences and based on information gathered from the participants in the practices under study. Thus, it is more likely to avoid being misled by the influence of the informal and volatile character of the subject matter. I believe that a foreigner – provided that he or she respects the specific nature of Russia – can deliver a *thick description* (Geertz 1973) of social processes in Russia. With all the foreigner's shortcomings, he or she still has the advantage of problematizing things that are obvious to insiders.

This study is written in English and by a scholar who comes from a country which, although it is a postsocialist one, is considered to belong to the West rather than to the East, especially in Russia. My goal was to provide a possibly deep understanding of that part of Russian transformation which was connected with petty trade. My long-term fieldwork in Russia, giving voice to my interview partners and Russian scholars were steps on the thorny path of achieving this goal. I cannot pledge my account is free of clichés or not biased in any way, but I can say that I have tried to look at the Russian situation through both internal and external lenses, as following only an internal perspective does not assure better results. Researchers who have investigated their own situations are also biased, though differently. Thus, it is the combination of various ways of looking that might lead to a deeper understanding of the phenomena studied.

As it turned out in the course of my fieldwork, many representations I had received from my earlier experiences and readings underwent revision. For instance, by the time I arrived, traders in Krasnoyarsk were not engaged in international shuttle trade any more. Urban open-air markets did not constitute a breeding ground for criminality and vendors were not harassed by racketeers as it was in the 1990s and as I had thought to still be the case. Finally, I was surprised by the number of labor migrants in Krasnoyarsk. At some open-air markets it was migrants who made up the vast majority of tradespeople.

These circumstances made me pose new research questions. I focused much more on the migrants and their relationships with the local population. I wondered what kind of identifications can be possible among people who 20 years ago used to be citizens of one state but because of historic changes became divided into locals and foreigners, hosts and intruders. Moreover, I was struck by the common use of the notions of *life* (*zhizn'*) and *survival* (*vyzhivaniye*). I started to explore this meaningful dichotomy and to explore what it means to local tradespeople and to newcomers. Last but not least, I turned to the investigation of so-called ethnic economy and posed the question of to what extent economic behavior is embedded in non-economic factors.

❖ The place

The decisive part of material for this book I gathered during qualitative ethnographic research among the people engaged in trade in Krasnoyarsk. Altogether, I spent almost 9 months there: from January to August 2012 (after staying 4 months at the HSE in Moscow) and from August to September 2013. There were a few reasons why I had chosen Krasnoyarsk for my research field. The information about Russia we get from the mass media usually only concerns Moscow, focusing particularly on the federal authorities and big business located there. However, it is very thought-provoking to look at Russia from the perspective of the Russian provinces, especially the ones that lie beyond the Urals. *Siberia is a colony of Moscow* or *they send us governors from Europe*, such statements are not uncommon in Krasnoyarsk and other Siberian regions and they show that the Asian part of Russia actually differs a lot from the European one. My research site, however, though more provincial than Moscow, had to be a city because trade is by its nature an urban phenomenon.

Krasnoyarsk has its own specific quality: the city developed very quickly during World War II. In 1941, when Germany attacked the Soviet Union, many factories were evacuated from the European part of the USSR to Siberia. It is at that time that the city became a major industrial center. Because of the presence of strategic (military, space, and nuclear industry) factories, Krasnoyarsk was a so-called closed city until the time of perestroika. That meant the city was under special surveillance and traffic to and out of it was subject to various restrictions. Much of the heavy industry located within the agglomeration borders, including chemical, metallurgical, and engineering factories especially, still operates. There is also RUSAL, the biggest aluminum plant in Russia.

Nevertheless, the 1990s brought grave changes to the city's labor market. Many plants were declared bankrupt and others were privatized, resulting in dramatic employment cuts. This was another reason why I chose Krasnoyarsk for my fieldwork. I felt that declassation would be particularly felt in a city known for its industries and inhabited in the Soviet times by a strong working class. This turned out to be true. As I later frequently heard from tradespeople in Krasnoyarsk, *nothing is produced but everybody trades* (*nichego ne proizvodit'sya no vse torguyut*). Many of them miss the Soviet past and this strongly affects their attitude towards their present work. They miss the past because as workers or engineers they enjoyed higher social status during socialism. In addition, many people (especially men) look at their previous workplaces with a sense of longing because they do not consider trade to be labor. *"If I worked at a factory, was socially useful and got paid regularly – this would be a true labor,"* I was told by Fedya (T9). Fedya valued his previous job – being a plumber in a factory – much higher than trading at a Krasnoyarsk open-air market.

The tradespeople's attitude to their job activities notwithstanding, Krasnoyarsk serves as an important economic and logistical center. With over a million inhabitants, Krasnoyarsk is the second-largest city in Central Siberia, after Novosibirsk.[48] The Trans-Siberian Railway and trans-Russian truck traffic cut through the city, which makes it a crucial road and rail hub. Krasnoyarsk is an important destination for labor migrants from the Caucasus, Central Asia, and China. It is also a supply source for the Russian Far North – mainly Norilsk – and neighboring regions of Krasnoyarsk Krai, such as Khakassia and Tuva. This gave me the opportunity to carry out comparative research among the wholesale traders from whom petty traders supply themselves.

There were also two minor reasons why I went to Krasnoyarsk for my fieldwork. Firstly, the city has been an anthropological *terra incognita*; according to my knowledge, no ethnographic study about the city has been written yet. The second reason was that I had friends in Krasnoyarsk. Such a circumstance is something important for an anthropologist, for whom social networks are both a subject matter and a key methodological tool. As a matter of fact, particular research sites are quite often chosen by anthropologists due to personal ties. Apart from theoretical considerations, especially at initial stages of the research process, it is simply very helpful if somebody shows you the surroundings, introduces you to other people, or puts you up for a few days.

48 It was in April 2012, i.e. during my stay in Krasnoyarsk, when the millionth inhabitant of Krasnoyarsk was born. The fact was broadly commented on in both local and the national press. Being million-strong enables a Russian city to get higher subsidies for transport infrastructure from the federal budget. Only such metropolises are also considered for building metro lines and – as I was told – considered in the corporate strategy of opening new McDonald's restaurants. Krasnoyarsk became the fourteenth, and easternmost, "million city" in Russia (cf. Golovanov 2012: online; Zyblev 2012: online).

❖ The people and the process

I conducted my fieldwork among people engaged in the street economy in Krasnoyarsk. My interview partners worked at open-air markets or sold informally on the streets, and they included both Russians and labor migrants from the Caucasus and Central Asia. A separate group were pensioners, who usually traded with produce from their gardens, which meant trade was not a steady work activity for them. In order to grasp the context of the investigated phenomenon, I also talked to people from outside petty trade in the strict sense, mostly to the wholesale traders from whom petty traders supply themselves. However, among my interview partners were also persons such as the owner of a bathroom equipment shop, a man who runs a garage with his brother, or an entrepreneur who offered lie-detecting services.

One of the peculiarities of street trade is that most of the people dealing with it are individual entrepreneurs. Moreover, these entrepreneurs usually work alone or with a family member, which means they do not employ any sellers and run just one trading point. Among all my interview partners who worked as traders, just a few possessed at least two stalls. Taking this into account, we can see that working in petty trade in today's Russia is a way of making ends meet rather than a prosperous business.

With some people I talked only once, and for just 5 minutes, while others I met several times and talked with for hours. An operational definition of a qualitative in-depth interview is thus called for here. I classify as an in-depth interview a conversation or series of conversation that lasted for a longer period of time and during which I got to know some essential information about a person: his or her career, the peculiars of their work, family situation, and so on. Altogether, then, I gathered around 80 more or less complete life stories. Apart from this, during the months I spent in Krasnoyarsk, I had literally hundreds of conversations – small and longer ones – with petty traders and other people who commented on their everyday life. After all, everybody has a household and everybody does shopping. Commonplace commentaries and observations constitute important auxiliary material for my dissertation.

Quite rarely was I invited to people's homes. Usually I talked to them at open-air markets or on the street. It was their workplace, then, which determined the possibility and conditions of holding an interview. On the one hand, it was up to me to adjust to every particular situation. For instance, I had to be careful not to block the view of someone's stall or in any other way disturb traders in their work. And when during a talk they got busy, they would simply stop paying attention to me. So the talks were often short and broken. On the other hand, however, this had the great advantage of allowing me to see the traders in their ordinary environment. I could experience their work conditions – I felt the cold or rain, for instance – and I could observe them in relations with other tradespeople and the customers. As a matter of fact, quite often, I would do just that, sitting behind a trader's stall, chatting with them, and simply observe what was going on in the alley.

Sometimes, however, my observation became participant. This was especially the case with two Azerbaijani brothers – Ibrahim (T17) and Kamran (T20) – with whom I established a friendship. I often helped them with their fruit and vegetables stand at the Slavynsky: I weighed the goods, served the customers, and so on. Sometimes I came to the market just to spend a few hours with them – it was a pleasure to sit and to talk with these two nice, open-minded, and honest people, who worked hard in order to secure the livelihood of their families. My friendship with the two brothers gave me an insight into dozens of small, and yet meaningful, situations which may happen at a market. I also could participate in the life taking place at the back of their stall. In very basic conditions, we often prepared meals and ate together there. I went with them to wholesale markets to see where, how and from whom they supply themselves in goods.

Nevertheless, the fieldwork was also quite tough for me, especially in the beginning. It was nearly -40°C when I landed in Krasnoyarsk. Within a minute after getting off the airplane I understood life there was different. My European winter jacket turned out to be laughable in Siberian conditions and I was forced to obtain another one, which I wore underneath. Later on, I learned from Kostya (V5) that cold is something you simply have to get accustomed to. Kostya belonged to a local winter swimming club and swam in the Yenisei River regardless of the weather.

As was mentioned, social ties are of great value for an anthropologist. It was through them that I met Tanya (T4), as she was acquainted with a friend of a friend. Tanya became one of my main interview partners and led me to the Slavyansky market at which she used to work for a long time. She introduced me to some of her friends but – to my disappointment – this was only partially effective. I did not know how to approach people and I felt they did not trust me. For reasons I was struggling to understand, either because of the cold or my lack of experience in contacts with traders, some people even did not want to talk to me.

Both my suspicions, concerning the cold and my lack of experience, turned out to be true – as the temperature outside grew, my market acculturation somehow also increased and it was much easier for me to establish relationships with people at the markets. I was also provided with a strong incentive to overcome the crisis when I met Fedya (T9). At that time, he was on the verge of bankruptcy and deeply depressed. His business, a fish stand at the Slavyansky, was going hopelessly. We spent hours talking outside his stall, in the container of his friend, or in the *stolovaya* (*bar*) at the market. Apparently, he needed someone to talk to. This was a lucky coincidence, since I badly needed someone to talk to, as well.

As Fedya revealed to me much later, besides the weather and my inexperience, there was another reason for my initial troubles. Some vendors from the Slavyansky thought I was a spy. *"They came to me and tried to persuade me not to talk to you. But I did not care,"* he said. A suspicious attitude towards foreigners is something you sometimes encounter in the post-Soviet space, especially in the parts in which people do not have much contact with newcomers from the West, which certainly dates back to the Cold War, if not to a more distant past. Moreover, in today's Russia, the attitude is only reinforced by the anti-Western (especially anti-American) rhetoric of the authorities and by legal regulations, such as the duty to register international NGOs as foreign agents. In addition, tradespeople have their own reasons for mistrusting strangers who come and ask questions. Although in recent years various inspections eased off at the Russian markets, in the mid-2000s traders were constantly harassed by different officials, who threatened them with fines and demanded bribes. Thus, I might have been perceived as a kind of inspector from a law-enforcement body, whether hygiene, taxation, or immigration.

Even knowing all this, it was a hard nut to crack for me to change people's attitudes. It is pretty difficult to prove that you do not steal if you are not a thief. In this context, meeting Fedya was important for me. Whether it was because of his contrary character or his business problems, but he really did not mind engaging in long discussions with me. I learned a lot from his story and his commentaries on the life at the Slavyansky and that deepened my sensitivity to tradespeople's problems. Furthermore, contacts with him helped me in psychological terms. They considerably boosted my self-confidence and sustained the belief that my project was accomplishable. Over the course of time, the Slavyansky became my main field of study and I managed to establish personal relationships with many vendors who worked there.

However, I frequently visited two other urban outdoor markets: the Tsentral'ny and the Krastets. Each of the markets had its own distinct character. While at the Slavyansky Russian and immigrant traders constituted comparable groups, the Tsentral'ny was occupied mainly by the former. The Tsentral'ny was the oldest market in the city. During socialism, it operated as a kolkhoz market selling foodstuffs. In the 1990s, it was significantly enhanced when clothing and home and repair sectors were added.

The Krastets, in turn, was a huge clothing market that appeared right after the fall of socialism and lay just within the city's limits. According to my interlocutors, in the 1990s the market served as the main source of clothes for the entire city and even today many of Krasnoyarsk citizens supplied themselves there. Local tradespeople were a slight minority there, as the majority was constituted by Kyrgyz, Chinese, and Tajik traders, although groups such as the Vietnamese or Chechens were present as well. The central parts of the Krastets were a place where good money could be made and that was why it was hardly possible to conduct real ethnography there. The narrow alleys were crowded with people, there was no possibility to stand in one place without blocking the view of someone's stall, and the competition was harsh, so the traders fought for customers and were never in a mood to devote their time to someone who just wanted to talk and not to buy. Thus, I preferred to walk around the less busy parts of the market, where possibilities of interaction were much greater.

Besides those three markets, I sometimes visited and talked to tradespeople from other city markets, just for the sake of comparison. I also talked to vendors who traded on the streets without any permits. The pensioners who sold their own produce were largely tolerated by the police, even if they did so by the entrances to supermarkets – which was quite often the case. Shop managers usually did not force them to leave, either, because it would make bad publicity for their shops. According to a common sense of justice, pensioners should be allowed to trade because they were supplementing their miserable allowances in this way. There were, however, also younger traders who tried to earn their living by trading informally on the streets. Those constantly risked being fined. I managed to get to know and to talk to a few of them, too.

I recorded only a minor part of my conversations. I never recorded people without their express consent. It happened to me a few times, however, that I lost the possibility to talk after I asked for permission to record. For the reasons I mentioned above, people were often afraid that I might misuse the recordings. That is why I often preferred not to ask and, consequently, not to record at all. Instead, immediately after coming back home from a market, I would write down in my diary everything I remembered from my conversations and observations. In addition, with some people I managed to establish a deeper relationship and several times we spent hours on discussing different matters. Even though these conversations were extremely informative for me, they were often of a private character and I felt it would be awkward to tape them. I noted these conversations in my diary as well, although I use them only to the extent that they were related to the issues discussed in this study.

❖ Positionality

One cannot liberate oneself completely from theory – just as pure facts do not exist, pure researchers do not exist, either. Likewise, one cannot disregard the fact of being of a particular nationality, cultural background, age, or gender. All these factors may influence the anthropological research process, both in a negative and a positive way. The risk of being considered a spy by some people can be counted among the shortcomings of my positionality in Russia. In general, however, my being foreign, Polish, and European constituted an advantage. Due to historical reasons, a substantial amount of Siberians have Polish ancestors. In addition, because of their common socialist past, many Russians are familiar with Poland. Indeed, Poland is considered to belong to the West – or to Europe in Asian parts of Russia – but it is a kind of West that it is not entirely strange. After all, both Russians and Poles are Slavs, which does manifest itself, if only in the similarity of their languages.

Apart from that, my origins had an ambiguous dimension in racial terms. Such terms are rarely expressed explicitly in Russia but they function in everyday relationships. Thus, ethnic Russians often treated me – as opposed to immigrant strangers – as *theirs* because I was Slavic, European, and white. Immigrants, in turn, considered me to be a stranger in Russia just like them. This was expressed explicitly by Fedya (T9): *"You and I – we will never feel at home here."* A few times I was also appreciated by immigrant traders because – to their surprise – I respected them and did not emphasize what they perceived to be my higher social status. In contacts with the Kyrgyz and Tajik traders, the experiences from my stays in those countries were very helpful. I knew the places from which the traders came, and this made it easier for me to establish relationships with them. My shortcoming in this respect, in turn, was that I did not command their native languages and had to communicate with them in Russian.

The dual dimension of my positionality was very useful to me, as representatives of each of the groups – locals and newcomers – usually did not hesitate to speak freely in my presence about members of the other. Personally, I empathized with both groups and tried to balance the amount of conversations I conducted with local and immigrant vendors. In general, however, the role of the ethnic factor in social relationships should not be exaggerated in any country whose history is as multiethnic as that of Russia. This applies above all to relationships among tradespeople, but it also applies to my relationships with my interview partners. There are situations that concern everybody. In facing dangers like the closure of a market, every trader is equal, whatever her ethnic or racial origins.

❖ Legitimizing ethnography

One could ask how one anthropologist can investigate petty trade in Russia. Even if one takes Krasnoyarsk alone, it seems unfeasible for a single researcher who uses qualitative methods to fathom the phenomenon of the urban street economy. There are over a dozen big open-air markets in this million-strong city with thousands of traders working there. There is also a massive amount of people who trade in the streets. Moreover, researching the historical development, economic dimensions, and social contexts of petty trade in contemporary Russia means simply aiming at researching a substantial part of the great transformation that has been taking place there (cf. Burawoy 2009: xv). So again, how can a single anthropologist do that? In asking this, we come to tackle the harsh question of the representativeness and legitimization of ethnographic research.

Even though moving between the micro- and macrolevels of analysis might seem risky, this is a quite natural motion, since "there can be no microprocesses without macroforces, nor macroforces without microprocesses. The question is how we deal with their relationship" (Burawoy 2009: 9). In fact, to link different levels of analysis and to look at a problem from different perspectives is intrinsically attached to participant observation. The observation "serves as shorthand for a continuous tacking between the *inside* and *outside* of events: on the one hand grasping the sense of specific occurrences and gestures empathetically, on the other stepping back to situate these meanings in wider contexts. Particular events thus acquire deeper or more general significance, structural rules, and so forth" (Clifford 1988: 34). Indeed, the ethnographic research process is made more complicated by the inescapable presence of the ethnographer in the world he or she studies. Ethnographers cannot pretend that they do not affect the reality in which they participate. Neither can they ignore the fact that a significant part of their sources are self-gathered. They come, ask, and make their interlocutors think and express things their interlocutors sometimes had not even thought about. These circumstances, however, must not paralyze the process of anthropological research.

On the contrary, "in our confusion is our strength" (Geertz 2000: 97). Bruno Latour (1993) argued that the somewhat manic passion to separate subjects from objects of research – that is, culture from nature – is not an ahistorical epistemological necessity but a phenomenon of modernity. The same applies to other rigid oppositions of positivist science: participant and observer, theory and experience, micro and macro, and finally, the boundaries between scientific disciplines. All these oppositions are more or less artificially constructed by cutting the flux of social world. The ethnographer who recognizes both the flux and his or her theoretical embeddedness in it can bring the poles of the oppositions into dialogue. This is the principle of reflexive social science (cf. Burawoy 2009: 8–9).

In the extended case method, developed by Burawoy, bridging the oppositions happens by four extensions: of observer, of observation, of microprocesses, and of theory. A researcher comes and joins the life of the community under study. To the extent enabled by the conditions and the researcher's skill, he or she penetrates into this life. Observations extend also over time and space. The researcher should stay in the field long enough to live with the community across various moments and to discern the rules and tendencies at play on-site. Apart from it, the researcher forges together the social microprocesses and macroforces, that is, he or she finds the macrofoundations of the former and the influence that the former have on the latter. Finally, the researcher should challenge his or her theory against the facts and experiences from the field and consequently improve that theory, until it becomes a theory that enables the researcher to turn the site into a case, that is, to go beyond the particularity of the site and to make it meaningful (cf. Burawoy 2009: xv, 17).

And that exactly is the way in which ethnography is legitimized. Indeed, ethnography does not fulfill the positivist requirement of representativeness – it cannot prove that the piece of the world under study is typical of the whole. The problem is, however, that neither can positivist science do so, and that we cannot even be sure at all whether the world consists of such typical pieces. "The search for an Absolute Truth is at odds with the relativist experience of anthropology" (Hastrup 1995: 50), and it is a mirage of such truth that stays behind the striving for representativeness as well as other requirements of positivist science. What reflexive anthropology proposes instead is constant reconstruction and revision of theory (cf. Burawoy 2009: 43–44, 53). Effects of such reconstruction are hardly expressible in numbers, but they are grounded in the fieldwork experience and their reference to social processes on the macroscale is cogent. It is the understanding brought by ethnography that serves as its ultimate legitimization.

Appendix 2. List of Interview Partners

❖ Petty traders: entrepreneurs and salespeople

T1 Galina, female, Russian, aged 44, shop salesperson at market of construction materials in Lyublino district, Moscow, 14 Dec 2011.

T2 Female, Russian, aged around 50, sells her own produce (pickle cucumbers, sauerkraut, etc.) informally in the street by underground station, Moscow, 15 Dec 2011.

T3 Pyotr, male, Russian, aged around 65, sells his own produce (honey) on the street informally by underground station, Moscow, 15 Dec 2011.

T4 Tanya, female, Russian, aged around 50, worked several years at the Slavyansky market and then in the Krastets market. Several conversations, first 2 Feb 2012.[49]

T5 Nadya, female, Russian, aged around 50, entrepreneur. Trades in fur coats at the Krastets market. 5 Feb 2012.

T6 Yekaterina (Galia), female, Russian, aged around 45, entrepreneur. Several conversations, first 6 Feb 2012.

T7 Natasha, female, Russian, aged around 35, salesperson. 27 Feb 2012.

T8 Talant, male, Kyrgyz, aged 24, salesperson. 27 Feb 2012.

T9 Fedya, male, Azerbaijani, aged 43, entrepreneur. Several conversations, first 28 Feb 2012.

T10 Sveta, female, Azerbaijani, aged around 40, entrepreneur. Several conversations, first 29 Feb 2012.

T11 Tanya, female, Russian, aged 28, salesperson at the Slavyansky market. Several conversations, first 2 Mar 2012.

T12 Alik, male, Azerbaijani, aged 42, entrepreneur. He used to sell fruits at the Slavyansky market, then he became engaged in petty wholesale trade in male underwear at the Krastets market. Several conversations, first 2 Mar 2012.

T13 Kyrlychbek, aged around 50, Kubanych, aged 28 and Mayrambek, aged 18, all male, all Kyrgyz, entrepreneurs. 6 Mar 2012.

T14 Zinaida, female, Russian, aged around 55, entrepreneur. 8 Mar 2012.

T15 Dzhargyl, aged around 30 and Ak Telek, aged around 35, both male, Kyrgyz, entrepreneurs. 8 Mar 2012.

T16 Two women, pensioners, Russians, aged around 65 and 75, sell produce from their dachas at social places. 10 Mar 2012.

---

49 Unless the place of interview is otherwise indicated, it took place in Krasnoyarsk.

T17 Ibrahim-Huseyn, male, Azerbaijani, aged 43, entrepreneur. Slavyansky market. Several conversations, first 15 Mar 2012.

T18 Rushat, male, Azerbaijani, aged 33, entrepreneur. Krastets. 22 Mar 2012.

T19 Tolik, male, Vietnamese, aged around 30, entrepreneur. Krastets. 22 Mar 2012.

T20 Kamran-Mikail, male, Azerbaijani, aged 40, salesperson, brother of T18. Slavyansky market. Several conversations, first 18 Apr 2012.

T21 Asylbek, male, Kyrgyz, aged around 45, entrepreneur. Krastets. 19 Apr 2012.

T22 Female, Russian, aged around 60, entrepreneur. Krastets. 19 Apr 2012.

T23 Murat, aged 35 and Nurdzhigit, aged 29, entrepreneurs, both male and Kyrgyz. 30 Apr 2012.

T24 Female, Russian, aged 70, pensioner, sells produce from her dacha at social place in the Slavyansky market. 3 May 2012.

T25 Mikhail, male, Russian, aged 70, sells potatoes from his dacha at social place in the Slavyansky market. 3 May 2012.

T26 Sergey, male, Russian, aged around 50, entrepreneur. Trades in fruits and vegetables at the Slavyansky market. Several conversations, first 8 May 2012.

T27 Tolik, male, Tajik, aged around 35, salesperson. Sells shoes in the Krastets market. 10 May 2012.

T28 Female, Russian, aged around 50, salesperson. Sells jackets in the Krastets market. 10 May 2012.

T29 Farahim, male, Azerbaijani, aged 25, salesperson. Cousin of T17. Several conversations, first 13 May 2012.

T30 Sulayman, male, Kyrgyz from Tajikistan, aged around 50, entrepreneur. Several conversations, first 14 May 2012.

T31 Adiljon, male, Uzbek, aged 28, entrepreneur. 15 May 2012.

T32 Alyona, female, Russian, aged around 60, entrepreneur. Trades in tea and coffee at the Slavyansky market. 16 May 2012.

T33 Misha, male, Tajik, aged around 40, entrepreneur. Trades in dried fruits and nuts at the Slavyansky market. 21 May 2012.

T34 Nikolay, male Russian, aged 55, entrepreneur. Trades in caps in the Krastets market. 22 May 2012.

T35 Female, Russian, aged around 65, entrepreneur. Trades in female underwear from Belarus. Krastets market. 23 May 2012.

T36 Vladimir, male, Russian, aged around 65, pensioner. Trades in second-hand things on a plastic cloth placed on the ground. Krastets market. 24 May 2012.

T37 Gayane, female, Azerbaijani, aged around 40, entrepreneur. Trades in fruits and vegetables in the Slavyansky market. Her husband sells informally potatoes by the entrance to the market. 25 May 2012.

T38 Feyrudin, male, Tajik, aged around 40, entrepreneur. Trades in dried fruits and nuts in the Slavyansky market. Several conversations, first 25 May 2012.

T39 Female, Russian, aged around 45, salesperson and her boss, female, Russian, aged around 25, entrepreneur. Both women are from Kansk, sell periodically seedlings in the Central market. 26 May 2012.

T40 Denis, Russian, aged around 25, owns a grocery shop, entrepreneur. 28 May 2012.

T41 Male, Russian, aged 74, pensioner. Sells own produce from his dacha in the Cheryomushki market. Formerly top engineer in a factory of nonferrous metals. 28 May 2012.

T42 Natasha, female, Russian, aged around 50. Trades informally in the street in pets (puppies and cats). 3 June 2012.

T43 Lena, female, Gypsy, aged around 16, salesperson. Trades in male clothes in the Central market. Works for her father who trades next to her. 4 June 2012.

T44 Valentina, female, Russian, aged 80, pensioner. She embroiders tablecloths, doilies, covers, etc. and sells them informally on the street. 5 June 2012.

T45 Oksana, female, Russian, aged around 45 and her son Roma, aged 23, entrepreneurs. Trade in meat and honey in the Central market. 6 June 2012.

T46 Misha, male, Tajik, aged around 50, entrepreneur. Trades in dried fruits and nuts in the Slavyansky market. 7 June 2012.

T47 Aleksandr, male, Russian, aged around 45, entrepreneur. Trades in domestic detergents in the Slavyansky market. Also does renovations. Several conversations, first 7 June 2012.

T48 Zhenia, male, Russian, aged 59, entrepreneur. Trades in mushrooms and berries gathered by him in the taiga and also with produce from his dacha. Central market, 9 June 2012.

T49 Irina, female, Russian, aged 55, entrepreneur. Trades in clothes in the Central market. 13 June 2012.

T50 Shafar, male, Tajik, aged around 35, entrepreneur. Trades in fruits and vegetables in the Slavyansky market. 14 June 2012.

T51 Lila, female, Russian, aged around 50, salesperson. Trades in fur coats in the Central market. 15 June 2012.

T52 Valentina and Pyotr, both Russians and aged around 50, entrepreneurs. Formerly *chelnoki.* They trade in jackets in the Central market and have two other shops in malls. Several conversations, first 15 June 2012.

T53 Zhenia, male, Russian, aged around 25, entrepreneur and Sasha, aged 27, formerly worked for Zhenia. Zhenia runs a shop with mobile phones and electronic equipment in the Radio market. 19 June 2012.

T54 Misha, male, Azerbaijani, aged around 50, entrepreneur. Trades in fruits and vegetables in the Slavyansky market. Employs one person. 21 June 2012.

T55 Anna, female, Russian, aged 76, pensioner. Trades in socks and knit sponges made partially by herself. Trades informally in the street. 22 June 2012.

T56 Dasha, female, Russian, aged around 55 and Sveta, female, Russian, aged around 55 – pensioners and entrepreneurs. They trade with cloths in the Central market. 20 Aug 2012.

T57 Sasha, male, Armenian from Georgia, aged 47, entrepreneur. Trades in male clothes in the Krastets market. 21 Aug 2012.

T58 Lubov I., female, Russian, aged 60, salesperson. Sales jackets and fur cloaks in the "Mega" shopping mall, next to Krastets market. 21 Aug 2012.

T59 Lyuba V., female, Russian, aged around 55, entrepreneur. Trades in groats, instant soups, macaroni, etc. in the Slavyansky market. Several conversations, first 22 Aug 2012.

T60 Galina, female, Russian, aged 82 and Yekaterina, female, Russian, aged 84, pensioners. They trade informally with produce from their dachas, they stay in the street by the entrance to a supermarket. 24 Aug 2012.

T61 Female, Russian, aged around 60, entrepreneur. Trades in biscuits in the Slavyansky market. 6 Sep 2013.

T62 Valentina, Russian, aged around 55, entrepreneur, trades in groats, rice, instant soups, tees and coffees, etc. in the Slavyansky market, 7 Sep 2013.

❖ Other business people

B1 Dima, male, Russian, aged around 40, owns a shop with bath equipment. 10 Feb 2012.

B2 Konstantin, male, Russian, aged 57, husband of Yuliya (B3). In socialist times he was a *shabashnik* and a *fartsovschik*. In the late 1990s he founded a big construction cooperative. Currently runs a company which supplies industry in the Far Russian North with specialist equipment. Several conversations, first 10 Mar 2012.

B3 Yuliya, female, Russian, aged 54, wife of Konstantin (B2). In the socialist era she ran a cooperative with Konstantin. Today she is employed in the company of Konstantin. Several conversations, first 25 Mar 2012.

B4 Valery, male, Russian, aged around 35, runs a shop with car parts with his friend. 18 June 2012.

B5 Yevgeny, male, Russian, aged around 45, owns a few shops and does consulting services. Two conversations: 18 June 2012 and 11 Sep 2013.

B6 Khudaybula, male, Tajik, aged 32, wholesale trader. Trades in onion and garlic in the Luch wholesale market, petty traders supply themselves from him. 4 Sep 2013.

B7 Three Azerbaijani men: Alik, aged 45, Halay, aged 43, and Altay, aged 42, petty wholesale traders. They trade together in fruits in the Juzhny wholesale market. 5 Sep 2013.

B8 Ravshan (Azror), male, Tajik, aged 60, wholesale trader. Trades in vegetables, mainly onions and potatoes in the Juzhny wholesale market. Two conversations 9 and 17 Sep 2013.

B9 Shakir, male, Tajik, aged 45, wholesale trader. Trades in onion, potatoes, and watermelons in the Juzhny wholesale market. Several conversations, first 10 Sep 2013.

B10 Lena, female, Russian, aged around 45, owns a grocery shop and a beauty salon. B4 also took part in the conversation. 11 Sep 2013.

❖ Various

V1 Aleksandr, male, Russian, aged 34, salesperson in a furniture shop, formerly academic teacher. Several conversations, first 7 Feb 2012.

V2 Grisha, male, aged 32, entrepreneur, does lie-detecting services. Several conversations, first 17 Feb 2012.

V3 Dima and Ira. Dima, male, Russian, aged 28, PhD student, computer scientist – works for Russian Railways – and entrepreneur: runs a construction company together with Ira, female, Russian, aged 26. 18 Feb 2012.

V4 Aleksandr, male, Russian, aged 33, entrepreneur (runs a garage with his brother). Several conversations, first 29 Feb 2012.

V5 Konstantin, male, Russian, aged 31, deputy manager and the salesperson in a company which produces honey. Several conversations, first 2 Mar 2012.

V6 Kirill, male, Russian, aged around 65, pensioner. He worked at the Krasmash factory and then he informally sold fridges. In the conversation also his wife Mariya took part, Russia, aged around 65, pensioner, and his son Maksim, aged around 35, employed in a private company, 4 Mar 2012.

V7 Tanya, female, Russian, aged 28, PhD student. In the 1990s she used to produce moonshine with her mother and grandmother. Barnaul, 29 June 2012.

V8 Aleksandr Pozdeyev, male, Russian, aged 31, academic teacher and entrepreneur. Two conversations 30 Apr 2012 and 18 Sep 2013.

V9 Andrey, male, Russian, aged 37, computer scientist. Several conversations, first 1 June 2012.

V10 Volodia, male, Russian, aged 37, Wrigley's sales representative, formerly represented various companies which sold vodka, tea, and coffee and delivered these goods to petty traders. 16 June 2012.

V11 Yury, male, Russian, aged around 50, and Vova, male, Russian, aged around 40. Yury works as a plumber and locksmith, formerly worked in Krasnoyarsk harvester factory. He is also the husband of a petty trader who works at the Tsentral'ny market. Wowa works as a bodyguard at this market. Formerly he was an engine driver for Russian Railways but partially lost his hearing and was forced to leave the job. 20 August 2012.

V12 Prof. Vladimir Datsyshen, male, Russian, aged 48, university professor (historian and sinologist). 3 July 2012.

V13 Ivan, male, Russian with Bulgarian roots, aged around 35, academic teacher (economist). Formerly deputy director of the city's business incubator. 27 Aug 2012.

Appendix 3. Voices from open-air markets

❖ Galina: *They work in order to survive, I work in order to live*

It's outsiders who work at the market; myself – I am from around here. I live, I go to the cinema, I visit cafes. They have to survive. They came here to work from China, Yerevan or some other country. These people need to survive. To rent a place to live, send home some money to someone, scrape something together for accommodation and other needs. I have everything: a car, husband, children, a flat. And so I do meet people who are envious. But that's just their lives. They work in order to survive, I work in order to live. I enjoy life. Their life is just work-sleep-work.

My official salary is 14,000 rubles. What's 14,000 in today's economy? It gets spent instantly to pay for the flat. But money is only my second or third priority. Of course I could do with more money, it's a natural wish. But what I find more important is the human factor. Where I work, the collective is like family. In my collective I have a colleague with whom I work. The CEO, the accounting department, managers, loaders, and warehousemen – I seldom see them all at once, only at joint events. The collective I worked in before, though – people were like wolves. I don't want it this way. Now I go to work with a peaceful mind.

And you can earn some money on the side as well. We do have a computerized cash register at the shop, but I can bring in my own merchandise and sell that – and the remainder of the retail price stays in my pocket. The CEO knows I do this. It's the market, it's what everybody does. Plus he understands no one is interested in working for 14,000. We're a limited liability company, so the pay is just basic salary, no bonuses. And for people to work well they need to be paid well. I do not steal. I do manage to make some money but not always. But that's commerce for you: feast today – starve tomorrow.

A construction worker comes in and wants to earn something off the supplies, say, he buys a bucket for 1,000 rubles and asks me to write it cost 2,000 rubles. So I write that, I let the man earn a living. If his client does not like the worker, considers him too expensive, he should buy his supplies himself – this way he'll know what costs what. This man comes to the market to earn money. And why not? When I redecorate my flat, I buy everything myself. But this requires going places, looking for things, it costs you fuel. The client has everything brought to him and so he should pay up – that's my opinion as well.

I don't think what I do is wrong. Everyone lives the life he or she can, and has the ability to, live. Myself, I would not be able to survive otherwise. Just imagine earning 14,000 rubles with two children! One daughter attends a fee-paying, the other – a non-fee-paying school. You need to provide for their education, put shoes on their feet and clothes on their backs. My only way of surviving is selling on the side. Those working in manufacturing, like my husband, have no such possibilities, all they get is the basic salary. Some work two or three jobs. All they ever do is work, twenty-four hours a day. When you have children, you have a reason to slave away like that. You have to get by somehow.

❖ Tanya: *Trade is a drug*

In the 1990s, everyone rushed to the market, out into the frost. The cold was very difficult to get accustomed to. Many, including myself, developed an allergy to frost. It's only now that they allow putting up all kinds of wind barriers or roofs, before we would just stand in the street. And it was fine, we all survived all right. What else can you do?

I sold newspapers. At first I worked for other news vendors, out in the street. But then they cheated me and I left. A lot of people would get cheated. They took girls from the street, set them up, and later refused to pay their wages. They came up with all sorts of reasons not to pay them. I left and started trading for myself. You find your suppliers and sell on. You look into what sells in newsstands and sell that. It was all right, it was swell, I liked it. You meet people, the day flies by, it's cheerful. The summers were of course easier and the winters – tougher. The temperature dropped to minus 30–35 degrees. But, you know, regardless of the temperature you had to get up at 7 o'clock and go to the market. Every morning you had to come and buy up the merchandise, it's a must, regardless of how frosty it was.

Before that I worked in a reading room at a library. Newspapers are my thing. And so I started selling newspapers at the Slavyansky market. The problem was that there was already a girl trading in newspapers at the market and I made her angry. She began sabotaging me, lowering her prices. She received her newspapers straight from the printing house, the employees printed them at night and then brought them to her at half the price. In other words, the newspapers were stolen. I had no such connections and had to supply myself through official suppliers, pay a big commission. And so I couldn't do normal trade and everyone went to her. I was of course mad. Then, gradually, I also found people who brought me [newspapers] unofficially. And I could earn a little bit extra.

Before, though, things were not going well at all. I could not switch to sausages, I could not switch to fish, nor to cheese. I wanted to, I could see people were buying those products, but to trade in them you had to cheat the customers. Those sellers live very well because they short-weight. I couldn't short-weight. I didn't go to church but I knew it was a grave sin to cheat people and that it would come back to bite me. I had my conscience.

We were cultured people working at the Slavyansky, everyone with a university degree. There are even professors working there to this day. There are people who are still unable to break free. Trade is a drug. It's a drug difficult to break free from. It's a drug in that the money you collect from your merchandise you have to put into merchandise again. And again. You're simply afraid to stop because then the money will end, the money will no longer come. There are no other jobs in Krasnoyarsk. You can pile up decent money from trade. I bought myself a house, a dacha, I went to see Prague. But you're simply afraid to quit, afraid to even take a break – they might take your place, take your merchandise away from you, your competitors might overtake you. And then you're left with nothing.

I liked it better when I worked for myself, nobody criticized me, I was my own boss. Then I moved to another market to work for a man. He was Vietnamese and had me sell padded jackets. He stood next to me when I traded. I found this very difficult, I found I was no longer able to work for someone else. I can only trade, work for myself. Whenever he would go away, I felt at ease and then it was all right. It's tough being controlled, I don't tolerate it. That's the kind of person I am.

I wish I could go back to that time. Because I was free then, nobody told me what to do. The customers would come, share their problems and I comforted them. After I left the market, they would remember me dearly and they remember me to this day. And we still go back with other traders to those days, how we helped one another. We always supported one another. The customers felt sorry for us, they would come and greet us on New Year's Eve and all the holidays. We lived our lives to the fullest. There really were few things that were bad. And we didn't get sick. I used to hurt all over and often catch colds – then I started trading at the market and never got sick.

❖ Nadya: *The state just abandoned us like a bunch of blind kittens*

The firm I'd worked for went out of business and we simply had no alternative but to go into commerce. Granted, I did graduate from a college of commerce, I am a trader by education, but I never imagined I would go into this line of work. But that's what life had in store for me. I thought at the time, "Well, all right. I'll do this for a year, two or three at the most." That was in 1993 and then it dragged on and on. At first I worked at home – I didn't go to the market, I sewed my goods at home. It was only in 1995 or 1996 that I began to trade at the market.

So I was sewing and people started having all sorts of remarks: "That's how you should do this, this is how you do that!" And I had to find out what the customers wanted from the customers themselves. You have to keep yourself informed about it all. The time of perestroika taught us to adapt... Before that we would not think too much about such things. There were designers, engineers, modelers working these things out. Now, after the state just let it all be, people are forced to look into all this themselves. They have to be designers, tailors, and to some extent engineers all at the same time, and even dig into the chemical side of things: the minks just get killed and brought to us and it's our business to "bring them to shape".

Take any given line of production – there are always many subtleties. As they say: "If you want to teach a man to swim, do not save him". It's the same here – if you want to teach a man to earn money, do not feed him, let him stay hungry and maybe then he'll achieve something. That's precisely what happened to us. As far as the state goes, well, it just abandoned us like a bunch of blind kittens. Nobody cared about us not receiving our salaries... Everything we produced in the [Soviet] Union – it all became useless and it all disappeared in an instant. And it was so difficult you cannot imagine...

In 1992 I went to Poland. Everything was just beginning there as well, and it was going the same direction. I would take with me, say, small mechanical equipment, tools, nickel silver and crystal products, and also silverware, all natural alloys. These things were in high demand. And we, in our turn, had a problem with clothes, so I brought back clothes. We didn't value what we had at that time. We ate from silverware, now it's plastic. We had gilded tableware, all those spoons, forks, cups. All this is much more expensive now. But we wanted a taste of the West – and so we brought back who knows what.

They would kill people then, all those criminal groupings. You were literally afraid you would lose your life. The very moment you start anything, any kind of business, they would start watching you. And it was not the state that controlled it. It was simply terror, terror and violence. You had to pay off racketeers. If now we pay taxes to the state – thank God that it's all civilized, it's all without fear. I go to the Tax Inspection Office and pay my dues for the three months I worked. There's no stress involved – this is simply your job and you settle your business's liabilities. But the times when trade had just started, when it was in its earliest stage in the 1990s – those times were terrifying. If you refused to pay your toll, they'd come and warn you a first time, they'd come and warn you a second time. The third time they would set fire to your shop and the goods. Or it could get even worse – there were murders, you know. That was horrible, barbaric. After 2000 it somehow got more peaceful. There were murders but only at the top levels. But the 1990s, until 2000 – this was a horrible time. We were constantly afraid. The 1990s were a time of terrible chaos [bespredel], there was no one to protect you. Police officers would leave the force because their pay was miserable.

Now it's gotten peaceful. Just pay for your spot and trade. But it's expensive. A 5-tonner costs 1,500,000 rubles – and that's just to buy the spot and trade there. No one can guarantee this container of yours will not be moved to a whole different spot, though. And there's no guarantee that anyone will come to shop at this new spot and you won't end up selling nothing at all. Not to mention that you cannot be sure whether or not you will manage to earn back those $50,000 – it might take you 10 years!

And it is not certain that they will not do away with the market. I don't know, it's only in our Russia that they can sell air to you. I put it this way because there's no certainty, there are no guarantees at all! They sell you the spot and the very next day come back and say: "We're sorry but the city administration plans to put some pipeline through this place and you are evicted." So there are no guarantees in this respect... And I cannot say the order at the market is ideal, either. The more money you give, the better your spot. Everything is based on money all the same...

We work in the frost. We work until it drops below minus 35 degrees because when the temperature is lower, it is simply impossible to stand for long periods of time. It's a stationary job after all. And there are no facilities to help you with that – what facilities can you have at an open-air market? Just imagine those temperature swings – from plus 35 degrees in the summer to minus 40 in the winter.

We dress very warm. Natural clothes only. I'm wearing a padded shawl and a padded jacket on top of that. It's no use putting on a mink fur – the wind is so harsh you'll be cold regardless. And you cannot wear a sheepskin coat – it won't keep you warm, either. You have to dress in modern clothing: either wind-resistant nylon jackets or natural-down-filled padded jackets – just make sure it stops the wind from blowing through. Now they've put a common roof above us to protect us from rain and snow. But nonetheless I doubt you can set our youth today to such work.

Recently a decree by Russian President Putin was published to do away with spontaneous [stikniynye] markets by 2013. I don't know what it will look like in reality – but I think that as far as Russia goes, this is simply impossible! These potatoes and carrots here – they were grown by local residents in their dacha gardens. And really... I'll always prefer this old lady, I'd rather buy her stuff than what the Chinese grow here, all fertilized with nitrogen. The old lady goes to a market in her neighborhood – and it's also convenient for me: down a few flights of stairs and I'm here to buy from her. That's why I'm more than certain the decree will fail.

The reasons they give for the decree is that such trade is not civilized. After all, we're entering the WTO. It's not modern that way, we need to build those huge shopping centers, seeing how Europe adopted this civilized course a long time ago. The basic premise is that people should be driven into huge structures that are heated, have the right conditions so that you could leave your children there as you would in the summer. To allow the children to just wander around a bit and make everything civil.

And one more thing – what's the tendency we've recently had? There are practically no Russians left at the markets today. It's mainly people from the "near abroad". Because their economic situation is much worse, much tougher, much more serious than ours. And they are a strong competition. You see – some of us had worked at the market for a very long time and many of them have by now left the street, because it's very tough to withstand such competition. Say you are able to earn 300–500 dollars, which in any case is not a lot by today's standards – you still have the Chinese, the Tajik, the Azerbaijani, the Kyrgyz, the Vietnamese willing to work for an even smaller salary because their economic situation is even tougher and there are no jobs when they come from. So there's hardly any of us Russians left at the markets. All Russian traders have found other jobs.

❖ Yekaterina: *You have to love your job*

China is bringing in this stuff now and it's really nice, too! We didn't have this in the Soviet Union. We didn't have this on our wedding nights – some of managed to find something in a shop, others got something from their elder sisters. And now women's underwear is such an interesting and varied merchandise. And the customers – young, interesting, bold! Sometimes funny, unpredictable... Well, now it's China itself bringing in all the goods. To go to China and fetch your merchandise yourself is not realistic anymore. The Chinese here take the goods straight from the factory all the way to the market! And they bring the stuff by boxcar loads! There's no way a bag-carrying Russian lady can compete with them. And even if I do go to China, what will I do there? I'll go to a Chinese market and bring my merchandise from there. I won't get into a factory, nobody will let me in, I'm not capable of that kind of output and that's why outplaying China is not realistic. The Chinese have boxcars of merchandise coming in, huge tonnage, so if I went to a market in China, the price I would get would in practice be the same as the one we get when we buy in bulk from the Chinese in Krasnoyarsk. So Russian women or sellers from Azerbaijan or Kyrgyzstan supply themselves here. The Chinese bring in huge batches and only sell in bulk – they're not interested in retail. What they want is to bring in as much as they can and sell as much as they can to Russian ladies like myself.

I wake up in the morning, put on thick trousers, a stupid cap and – no longer resembling a woman – go to spent the entire day out in the frost, in the cold. My arms hurt, but I'm used to it by now, I've learned to massage them. Of course it's tough but you get to love it. I've been at the market since 1990, that's 22 years now, and I love the market! The market is my thing! Every man should like his job, if you don't like your job, you will never be of any use doing it – if you just go to occupy space. You have to love your job! You need to love your merchandise, you have to love new merchandise! There's a new batch coming? Hoorah!

I had another profession – I am a certified confectionery technologist – and I thought that would be my line of work, but there's no leaving the market! Once you grow to love it, it sucks you in and there simply is no leaving then – it sucks you in so much, it's so interesting... And the money: if you get the drift of it, if you find it in you, you will have real money. But first of all, you need self-discipline, after all you have no bosses above you. I wake up in the morning, it's cold outside, minus 15 or minus 20 degrees, but I should get up, get dressed and "forward march!" to work. You won't get a note for being absent, no manager will tell you off – and still you spend your whole day standing at the market, from opening till closing time.

The first 12 years I sewed caps for the Chinese market at the Krastets. But caps don't sell all the time, only in the winter, and I needed something to live on during the summer. So I would sell my produce and then begin my search: what do I do? What should I sell? I had to think constantly, after all I also had children to feed. Simply put, I had to survive. Here, in Siberia, you won't be able to survive if you don't work – especially when you're a single woman with two children. There's no money and you have to work. So we went to Saratov to get our merchandise – we would bring in knitwear: socks, mittens, also bask wisps and all sorts of aprons and gowns. Then we went to Moscow as well. I never went to China, but I've been to Khabarovsk, Blagoveshchensk, Irkutsk, Chita...

I started selling underwear quite a long time ago, it's been 10 years. We used to trade in socks, and they sold well. I liked this merchandise. But as it turned out, one of our competitors was given a lot of spots by the market administration – and one of those spots was next to ours. And so she would watch the merchandise we would put out and she would copy what we did exactly. We put out checked socks, she did the same – only she had 6 outlets and our stuff hardly sold at all. And so we slowly started selling fewer and fewer socks, until we gave them up altogether and put out pants. There was nothing we could do about her – that's administration for you. Nothing anyone can do. We shut up and switched merchandise – and for all these years it's been going well, we work.

Those who cannot survive, leave. Some lack the brains, others the money. The brains are necessary – you bought the stuff, everybody is smiling at you, you've put your merchandise out... and now you need to sell, to earn back the money. Only then will you start earning your money on top of that. Not to mention that you have to pay up the administration, and that's expensive as well. And you can't bring them leftover panties and bras, can you? What I'm saying is, not everyone makes it. Not everyone can withstand the cold, not everyone has the health for it. Besides, there's no boss to tell you, "Sit back there!". So, many just pack their things and leave early. Many people start drinking vodka when it gets cold...

All kinds of people work at this market. The Chinese come, the Vietnamese, people from the South come here to feel the frost – the Tajiks, the Uzbeks, the Kyrgyz. We have diasporas emerging, that is, there's the head of the Chinese, the head of the Tajik, the head of the Vietnamese. There's no head of Russians. As a matter of fact, the market administration told us: "Well, we don't need Russians at the market, you can go eff yourselves." We've stood our place for so many years and now, as it turns out, chop-chop and our place is gone! "We can give you money." "No, no, we don't need Russians." And the reason they don't need Russians is that it's better for them to take a Kyrgyz girl and have her work for peanuts. Not to mention that us Russians also drink and are not very disciplined. But take a Kyrgyz girl, tell her to scrub the floor and she'll start at 7 a.m. sharp and work sparing no efforts, as they say. Not like our Russian do-nothings. Well, not all of them, of course.

Concerning the rising costs, there's the renting fee and then another fee – they call it "the guarantee fee". Its full name in official Russian is "Fee guaranteeing that your hitherto purchased place is not sold to another natural person". "Pay up!" In other words: "Buy the container or we'll sell it to another person." That is, you should buy this steel box – a container costs, say, 500,000 rubles, 1,200,000, 1,300,000 rubles, depending on the type, whether it's big or small; the fee for a big, 20-square-metre container is 60,000 rubles a month. That's the renting fee. And the guarantee fee is when they come every six months and say, "Pay up the guarantee." The guarantee makes an additional 60,000 rubles and you have pay it every six months. If you don't, they can sell your place to some other trader. Even if you've bought this same container beforehand: "Fee guaranteeing that your hitherto purchased place is not sold to another natural person".

And the 60,000 rubles does not include everything, either. For example, they produced plastic bags and other packaging with the market's logo on them and gave everyone 200 pieces each, demanding 600 rubles. That's the administration's way of cooking up some income. Then there's seals, price tags – if your merchandise does not have the administration's price tags on it, the market's security guard will fine you 500 rubles and keep the money for himself. That's the way we have it here, it's all fines, fines, fines! They won't take much from us Russians, but the Vietnamese do pay up. I mean, a guard just comes by and looks at him and the Vietnamese pays instantly. And it's no wonder, the Vietnamese are all small and our guards are all big guys.

They've been closing down our market for 15 years now... But the truth is, this market feeds the entire city administration. Just imagine getting 60,000 rubles for every container! Where does this money go? Part of it goes to the owner, part to the city administration, part to such-and-such, and everyone's all smiles. Everyone needs the market. Everyone wants money! So if you were to close it, what would the city live on? So yes, they'll close it all right – the word will spread – and then they'll open it all over again. They're given money and they open it again.

We do have competition. We have lots of Kyrgyz people. Sometimes so many of them swarm out... But why should I blame them? I don't mind if other people sell what I sell. They also have to live, to take care of their children. The woman I trade with and I are in a situation where we come, receive our money, pay the Chinese for the merchandise, split the remainder, and go home. The Kyrgyz are a hard-working nation and they come here to earn money. They're hard-working and strong. They are stronger than Russians, Tajiks, or Azerbaijanis. No one compares to the Kyrgyz. Very strong they are! A husband and wife come – very young, perhaps 20–22 years old. And they are brave, they start having children here. She works until the very last day of pregnancy, the cramps begin right here at the market but she just carries on and takes it. Later I look at her and she's all pale. So I tell her, "Anya, do you feel bad?" "No, I'm fine." "Anya, I can see you're not feeling well!" "No, I'm fine." Then I go, "Girls, she's in labor and she *'s just standing there working!" We call in the husband and say, "Call the ambulance right away! Take her to the hospital as fast as you can!" They barely made it in time...*

And so they have a child here, after a month or two they find a nanny, also from Kyrgyzstan – a sister of one of them or some other girl. She comes, they agree on five or six thousand rubles a month. This is the money they send back to her relatives in the country. The girl nurses the child, and lives and eats with them. The girl is fed, the family receives something, the child is looked after. So they come here from Kyrgyzstan in big groups and each shares a flat with 7–9 other people because the rent is substantial. And still they start having children right away!

Some send the children to their homeland for their grandmothers to watch, some bring them up here. Let's say the father and the mother are at the market – their little son, who's in the first grade, ought to come to the market after school and help his mother at the market... He's already helping, already working! If his mum sells tea, the little one will run to fetch her boiling water in thermoses. The thermoses as almost as big as he is; on his way back he's barely able to carry them. But at home they need to eat, they need money. And another child is helping the mother put out the merchandise at the market, and yet another, a 6-year-old, is at home watching over a 2-year-old.

We had one incident here. A Kyrgyz first brought his brother, then his sister, then he himself got married. Soon the wife was pregnant, it's her eighth month and she left the merchandise, she went away from her stand, leaving the merchandise behind... That evening he beat this pregnant woman so severely! We were simply shocked!... "Don't you dare leave our stuff!"

❖ Aleksandra: *People come to the market to talk*

They'll be closing this market. I think it might last another year or so. So we'll have to change something, we won't be able to just stick around here our entire lives. Well, that's what the state policy is. Any minute now state will come and they'll close it. No one is asking us, that's for sure. This is policy of the authorities: to get rid of small and midsize business, close down all kiosks and stalls, all the markets – and leave only supermarkets. Everyone is wondering what to do now.

For some time now there've been less and less customers at the market. As quiet as it is now... this is literally the first year when it's gotten so bad. There were never so many supermarkets, kiosks, and shops. They've opened too many in this area recently. They're building something new every year. Of course some of the customers leave to go there. It used to be impossible to just go around the market like that, to just take a walk. There was just a river of people, a real nightmare! People would go and go... Now even when we stand in the street and regular customers walk by, we ask them, "Why do you pass us by?" And they say, "It's the crisis. We really do have no money, cheese is not a necessity, we think we can live without it for now."

People got used to coming here, to the market. They come to talk, to complain about things, tell somebody about their problems. Another reason they come is that in a supermarket you can't even buy a 200-gram piece. It's all portioned, you have, say, 500- or 600-gram pieces. Whereas generally, people go to buy 200 grams for breakfast and that's that. So they come to the market and we'll cut off as much as they like. There's no way they'll have that at a supermarket.

Over the years, we've built up a group of regular customers. They come once and they like it that we talk with them, offer them advice. They say, "We came to you once and the cheese you gave us was so tasty, we liked it so much." They come once, they come twice, they come the third time, and finally they don't even say what cheese they want, they just ask, "Give us something according to your taste, what you like. You always give us good cheese."

We attract customers with our sympathy – everyone wants to have someone to complain to. We'll hear them and talk to them. As far as the Russians go, it's mainly women who work at the market. And when men come, they like it – and so we know everything about their grandmothers, grandfathers, children, grandchildren, they'll tell us about everyone. The truth of the matter is that you've never even seen these people in your life but you know all there is to know about them. That's why they come – to talk. We have such clients. We had one, he would always come but he was exceptionally silent. Until one day he came by drunk. And that was that! He kept talking for four hours straight, told us all his life. "Now I have someone to talk to, I can see you girls are good people..." and he just stood and stood there. Always the silent type, all he said was "good day, goodbye". And then, all of a sudden, a dam broke.

Of course we lack certainty. Our boss wanted to redo the kiosk before the winter, when we trade from inside through a small window. It's uncomfortable, we wanted to let the customers in, but that's impossible, there's not enough room. So he was planning to put up plastic walls, cover everything with plastic to close this space up and let the people in – they would be warm, it would all be clean, everybody would be happy. But then he said, "What's the point when I don't know whether or not the market will still be standing? I'll put money in it now and in three months they'll tell me, 'That's it, clear the area, tear it all down.'" Of course this lack of certainty bothers us. If they told us, "All right, you'll be here another three years," then for the sake of these three years, we could develop. The way things are now, we don't know whether or not to start the construction. Of course that's bad. But that's just how it is here... Nobody tells us, the common people, anything. They make some decisions among themselves and that's that. We have an administration that won't even talk to us. They'll just scold us and say, "It's not your business, go away." No one says anything.

❖ Nikolay: *Capitalism began and people stopped receiving their salaries*

More than half of factories in Krasnoyarsk are defunct. Under socialism industrialization levels were high, we were an industrial center, whereas now few factories operate. After all, they succeeded in China, they managed to reorient the economy. Well, here they just looted the factories and then sold them away. They closed down the tire factory, the industrial rubber plant, the silk factory, that is, you see, they were reoriented towards commerce.

I've been at the market since 1992. Originally, I worked at the silk factory, I was a factory worker. When the cooperatives kicked off, I joined one at first, they paid better. And then I went to do business on my own. In the early 1990s, factories stopped paying people, you know – capitalism began and people stopped receiving their salaries. The factories were choking with products which nobody wanted and which it was impossible to sell. It was then that I realized I had to eat and the market was the only way to go: either starve or go to the market.

Over these 20 years I've spent at the market some have risen, some have fallen, some have stayed where they were financially, some have left to work other jobs. Now it has gotten bad. The competition's serious. The Chinese... We used to have no Chinese. Russians brought stuff from China themselves and that was better, that was good. The factories worked, people had money. Then they stopped having money and what happened? The Chinese came. I bring something and sell it for 10 rubles, he brings the same and sells it for a ruble. Of course people will go to him, not to me. So what's the point?

The Russians went and we're left with the Chinese. The Chinese, the Kyrgyz, the Azerbaijanis... They arrange for citizenship and all the papers – with bribes, how else? Bribery is the blood of Russia. It's become the blood of Russia. There are no laws for that. Well, there are laws, but they're ineffective. Everyone's taking bribes.

Twenty, thirty years ago it was the same for everyone, everyone was equal. Now we're stratified in a big way. And why is that? Because of connections. It's not about any resourcefulness, it's about connections: they meet, make a connection and that's it, thank you very much. One of them takes the oil, sells it on, and keeps the money. Just steal and keep the money. There's no civilization here.

Russia is switching to a virtual economy. Everybody's just waiting for the worker to produce a bolt and a nut. Then they start selling this to one another, there are always plenty of intermediaries. And everyone takes a salary for it – it's all salaries, salaries, salaries. And when the tenth man in the row eventually sells to the state, it turns out the bolt and nut are golden, or at least they cost as much as gold. And that is why the state has no money. They simply come up with virtual transactions but end up receiving billions of real rubles. And then they buy real houses, real cars, real delicacies to eat. I don't think it's fair. If you do virtual work, you should eat virtual food.

But these are sly people operating, so they arranged it so that the rich pay no tax at all, the medium ones pay a little, but the taxes they made for the smallest ones are huge. They took from capitalism what suited the rich. But they didn't take... Well, decent wages, this they didn't take; they didn't take trade unions, the real ones which make an impact, not the obedient ones but ones that can bring people into the streets and that those in power would listen to. And what did they take? All that suits them they took, and whatever doesn't suit them: "Why, no, no, we don't need that kind of capitalism, we'll go our own way."

We should have adopted what they call the Swedish, the Finnish way. From the very beginning they offered Yeltsin, "Let's follow Sweden, Finland". But he said, "No, no, we'll go our own way." And the truth was that he wanted to steal a lot of money. And Yeltsin did steal a lot of money and the people who surrounded him stole a lot of money. They just took away factories – the non-ferrous metallurgy, the oil industry – they took it all for themselves and gave it away to their friends. And that was it, we're now a nation of beggars.

Now I have nowhere to go. And I have no idea what to do. And I have nobody I can cheat. And besides, I don't have the conscience to cheat. Otherwise you just swindle people and go to Spain. Look at what's happening: somebody steals money and they go to Spain, another one stole and went to London, a third one stole and off to Paris. Nowadays half of London are rich Russians: officials, thieves. Luzhkov, Abramovich, Berezovsky, all of those people are citizens of Great Britain. And we're left here and they stole from us and live a good life there.

❖ Yevgeny: *Today in Russia the only work is in oil, gold, diamonds, and... kiosks*

I've been in the business since 1990. In the 1990s, it was a very profitable thing to do because there were no tax inspections, just the administration – city district administration, that is. They would issue a permit and there would be practically no taxes to pay. Plus, as you remember, it was the time of perestroika. There were no goods in Russia, we would bring things from Poland, from everywhere. Actually, there was nothing at all, nothing to sell, but there where queues all around anyway. So everything we brought sold well and the profits were very high.

At the time, I sold groceries and spirits, this sort of thing. And sometimes after just a month of work the profits could buy you a big car like this [points to his Toyota RAV4]. *That's the kind of profits we had. Then year after year it started to go down. And now it sort of depends on the location, the place you trade – if trade is brisk, the place is good, the profit's sort of higher, and if you trade, sort of, on the outskirts of the city or in a place that is, sort of, not so good, then there's less profit. It still pays to run a small business, it still pays to, sort of, run kiosks now. Because now, in Russia there's practically nothing else to do for – I don't know if it's the right way to call it – the middle class. There are no manufacturing enterprises, and the few that we have are small ones. I used to work at an escalator factory myself.*

*You see, there used to be criminals all around. Just like that, everyone around would pay money to be able to trade. There was a racket all right. That's what we call criminals in Russian, "*reket*". Without this you would not be able to trade. The entire city was divided between groupings, and each grouping would collect money from the "*kommerses*", that's how they called people engaged in commerce in their criminal slang. And* *so they would collect money from everyone. Each shop, each stall, each factory, almost everyone paid. In the 90s, practically 90 percent of enterprises paid to be able to work, because the authorities were unable to secure the entrepreneurs' rights, to defend them. Only criminals were able to defend the entrepreneurs. These were the beginnings. Now they, sort of, brought back order, there are no criminals now, or at least there's very few, perhaps 1 percent, perhaps even less.*

Nowadays, the police have, sort of, started working in the right direction, because earlier the police would, sort of, work for the criminals. They were sort of cooperating. And now it's all right, the police are doing a good job. And owing to this it's now generally better, because there is less criminality, there are fewer violent acts, there are less people killed in the streets. They used to kill many people straight in the street. A man comes along carrying a machine gun – bratatat! – and that's it, the man is gone. Now it's sort of all right and interesting to work, and now the important, the most important thing is to let entrepreneurship develop and not choke it with regulation. All the entrepreneurs are people who want to work, and on a lawful basis at that.

And besides, there's simply no alternative in Russia today, nothing else to do. Many people would like to quit this business because it means working day and night: you have to stock up, you have to make sure the sellers are all right. You check all this and everything else. Business is problematic but at the moment there's nothing else you can do in Russia. Today in Russia the only work is in oil, gold, diamonds, and... kiosks.

The Slavyansky, the Kolkhozny, the Krastets are all very comfortably located and if you close these markets, the people, the inhabitants of the city will sort of... they'll simply pay much more for the groceries and the clothing they need. Nowadays, when you go to a supermarket, to buy the groceries for one meal you'll easily pay five or six thousand rubles. So, for a monthly salary you can visit the supermarket two times. If you take the same money to the market, you can come a few times and you'll get better quality. This means tearing down the markets only suits supermarket owners and that is the reason they want to tear down the kiosks. It's them who lobby for those laws.

[The selection of interviews was translated from Russian by Jakub Ozimek]

Works cited

Afanasyev, Yury N. (2001): Опасная Россия: Традиции самовластья сегодня. Istoriia i pamiat. Москва: РГГУ.

Aldrich, Howard E., and *Roger Waldinger* (1990): Ethnicity and Entrepreneurship. Annual Review of Sociology 16: 111–135.

Aleksandrov, Nikolay (2013): Кто ответит за базар? http://www.sgzt.com/krasnoyarsk/?module=articles&action=view&id=1918&theme=2, accessed March 1, 2014.

Aleksashenko, Sergey (2009): Битва за рубль: Взгляд участника событий. Москва: «Время».

Anderson, David G. (2000): Surrogate Currencies and the "wild Market" in Central Siberia. In The Vanishing Rouble: Barter Networks and Non-Monetary Transactions in Post-Soviet Societies. Paul Seabright, ed. Pp. 318–344. Cambridge: Cambridge University Press.

Arbitraznhny sud Krasnoyarskogo kraya (2013): Решение от 01 февраля 2013 года по делу № А33–15523/2012. http://docs.pravo.ru/document/view/29787667/30513511/, accessed March 1, 2014.

Aristotle (1988): Politics. C. D. C. Reeve, tran. Indianapolis/Cambridge: Hackett Publishing Company.

–: (2011): Nicomachean Ethics. Robert C. Bartlett and Susan D. Collins, trans. Chicago and London: The University of Chicago Press.

Ashwin, Sarah, and *Tatyana Lytkina* (2004): Men in Crisis in Russia: The Role of Domestic Marginalization. Gender & Society 18(2): 189–206.

Aslund, Anders (1991): Gorbachev's Struggle for Economic Reform. Updated and expanded ed. Studies in Soviet History and Society (Ithaca, N.Y.). Ithaca, N.Y: Cornell University Press.

–: (1995): How Russia Became a Market Economy. Washington, D.C: Brookings Institution Press.

–: (1997): Economic Causes of Crime in Russia. In The Rule of Law and Economic Reform in Russia. Jeffrey Sachs and Katharina Pistor, eds. Pp. 79–94. Boulder, Colorado: Westview Press.

–: (2002): Building Capitalism: The Transformation of the Former Soviet Bloc. Cambridge; New York: Cambridge University Press.

–: (2007a): Russia's Capitalist Revolution: Why Market Reform Succeeded and Democracy Failed? Washington, DC: Peterson Institute for International Economics.

–: (2007b): How Capitalism Was Built: The Transformation of Central and Eastern Europe, Russia, and Central Asia. Cambridge: Cambridge University Press.

Aven, Pyotr (2011): Петр Авен о том, как он и Альфред Кох искали правду о 90-х. http://lectures.gaidarfund.ru/articles/1149, accessed April 16, 2013.

Avilova, Agnesa V., Andrey O. Blinov, Andrey A. Voz'mitel', et al. (1999): Малое предпринимательство в кризисном обществе ("круглый стол"). СОЦИС 7: 78–90.

Avshalumova, Rimma (2013): Куда исчезает малый бизнес. http://www.vedomosti.ru/career/news/16314211/massovyj-ishod-v-neformaly, accessed August 20, 2014.

Barsukova, Svetlana (2004): Неформальная экономика: экономико-социологический анализ. Москва: Издательский дом ГУ-ВШЭ. http://publications.hse.ru/en/books/54925521, accessed October 9, 2013.

–: (2009): Неформальная экономика. Курс лекций. http://publications.hse.ru/en/books/54925848, accessed September 21, 2014.

Beck, Ulrich (1992): Risk Society: Towards a New Modernity. Theory, Culture & Society. London: Sage Publications.

–: (1999): World Risk Society. Cambridge, UK: Polity Press.

Beck, Ulrich, and *Elisabeth Beck-Gernsheim* (2002): Individualization: Institutionalized Individualism and Its Social and Political Consequences. London: SAGE.

Blyakher, Leonid E., and *Nikolay A. Pegin* (2010): Динамика представлений населения Дальнего Востока России о китайских мигрантах на рубеже XX–XXI веков (на материале интервью с предпринимателями). In Миграции и диаспоры в социокультурном, политическом и экономическом пространстве Сибири. Рубежи XIX–XX и XX–XXI веков. Viktor I. Dyatlov, ed. Pp. 485–501. Иркутск: Оттиск.

Boettke, Peter J. (1993): Why Perestroika Failed: The Politics and Economics of Socialist Transformation. London: Routledge.

Boholm, Åsa, Annette Henning, and *Amanda Krzyworzeka* (2013): Anthropology and Decision Making: An Introduction. Focaal—Journal of Global and Historical Anthropology(65): 97–113.

Bonacich, Edna (1973): A Theory of Middleman Minorities. American Sociological Review 38(5 (October)): 583–594.

Boym, Svetlana (1994): Common Places: Mythologies of Everyday Life in Russia. Cambridge, Mass: Harvard University Press.

Brandtstadter, S. (2007): Transitional Spaces: Postsocialism as a Cultural Process: Introduction. Critique of Anthropology 27(2): 131–145.

Brednikova, Ol'ga, and *Oleg Pachenkov* (2001): Азербайджанские торговцы в Петербурге: между «воображаемыми сообществами» и «первичными группами». Диаспоры(1): 131–147.

–: (2002): «Этническое предпринимательство» мигрантов и мифы мультикультурализма. In V. S. Makhalov and A. Tishkov, eds. Pp. 155–161., Мультикультурализм и трансформация постсоветских обществ. Москва: Институт этнологии и антропологии РАН.

Brooks, Karen M. (1988): The Law on Cooperatives, Retail Food Prices, and the Farm Financial Crises in the U.S.S.R. Staff Paper P88–29. St. Paul, Minnesota: University of Minnesota. Department of Agricultural and Applied Economics. http://ageconsearch.umn.edu/bitstream/13265/1/p88-29.pdf, accessed March 21, 2013.

Bruni, Luigino (2002): Toward an Economic Rationality "Capable of Communion." In The Economy of Communion. Toward a Multi-Dimensional Economic Culture. Luigino Bruni, ed. Lorna Gold, tran. Hyde Park, New York: New City Press.

Bryant, Christopher G. A., and *Edmund Mokrzycki, eds.* (1994): The New Great Transformation? Change and Continuity in East-Central Europe. London: Routledge.

Buchowski, Michał (2001): Rethinking Transformation: An Anthropological Perspective on Post-Socialism. Poznań: Humaniora.

–: (2004a): Hierarchies of Knowledge in Central-Eastern European Anthropology. Anthropology of East Europe Review 22(2): 5–14.

–: (2004b): Redefining Work in a Local Community in Poland. „Transformation" and Class, Culture, and Work. In Workers and Narratives of Survival in Europe. The Management of Precariousness at the End of the Twentieth Century. Angela Procoli, ed. Pp. 173–196. New York: State University of New York Press.

–: (2006): The Specter of Orientalism in Europe: From Exotic Other to Stigmatized Brother. Anthropological Quarterly 79(3): 463–482.

Bugakova, Yana (2013): Старейший рынок Красноярска "Славянский" закрывают. http://afontovo.ru/news/13619, accessed March 3, 2014.

Burawoy, Michael (1999): The Great Involution: Russia's Response to the Market. Retrieved December. https://brainmass.com/file/33548/Polanyi+From+Russian+Perspective.pdf, accessed September 3, 2014.

–: (2001): Transition without Transformation: Russia's Involutionary Road to Capitalism. East European Politics and Societies 15: 269–290.

–: (2009): The Extended Case Method: Four Countries, Four Decades, Four Great Transformations, and One Theoretical Tradition. Berkeley: University of California Press.

Burawoy, Michael, and *Katherine Verdery, eds.* (1999): Uncertain Transition: Ethnographies of Change in the Postsocialist World. Lanham, MD: Rowman & Littlefield.

Burchell, Graham, Colin Gordon, and *Peter Miller, eds.* (1991): The Foucault Effect: Studies in Governmentality: With Two Lectures by and an Interview with Michel Foucault. Chicago: The University of Chicago Press.

Bychkov, Valery (2013, 2014): исчезнут ли следующие 600000 ИП? http://smartsourcing.ru/blogs/buhgalteriya_i_finansy/2260, accessed August 20, 2014.

Caldwell, Melissa L. (2011): Dacha Idylls: Living Organically in Russia's Countryside. Berkeley: University of California Press.

Cantillon, Richard (2010) [1755]: An Essay on Economic Theory. Mark Thornton, ed. Chantal Saucier, tran. Auburn, Alabama: Ludwig von Mises Institute. http://mises.org/books/essay_on_economic_theory_cantillon.pdf, accessed April 12, 2013.

Chari, Sharad, and *Katherine Verdery* (2009): Thinking between the Posts: Postcolonialism, Postsocialism, and Ethnography after the Cold War. Comparative Studies in Society and History 51(01): 6.

Chernyshevsky, Nikolay Gavrilovich (1989) [1863]: What Is to Be Done? Michael R. Katz, tran. New York: Cornell University Press.

Chłopecki, Jerzy (1990): Państwowe przedsiębiorstwo w szarej gospodarce. In Gospodarka nieformalna: uwarunkowania lokalne i systemowe. Kazimierz Z. Sowa, ed. Pp. 109–234. Rzeszów: TNOiK.

Chwalba, Andrzej (1995): Imperium korupcji: w Rosji i w Królestwie Polskim w latach 1861–1917. Kraków: Universitas.

Cichomski, Mariusz (2010): Stadion Dziesięciolecia – „awangarda polskiego kapitalizmu". In Wielkie bazary warszawskie: środowisko społeczne, kultura i problem społeczny. Jacek Kurczewski, Mariusz Cichomski, and Krzysztof Wiliński, eds. Pp. 209–231. Warszawa: Trio.

Clarke, Simon (1992): Privatization and the Development of Capitalism in Russia. New Left Review 196: 3–28.

–: (1999): The Formation of a Labour Market in Russia. Cheltenham, UK; Northampton, MA: Edward Elgar.

–: (2000): The Household in a Non-Monetary Market Economy. In The Vanishing Rouble: Barter Networks and Non-Monetary Transactions in Post-Soviet Societies. Paul Seabright, ed. Pp. 176–206. Cambridge: Cambridge University Press.

–: (2002): Making Ends Meet in Contemporary Russia: Secondary Employment, Subsidiary Agriculture, and Social Networks. Cheltenham, UK; Northampton, MA: E. Elgar.

Clifford, James (1988): The Predicament of Culture: Twentieth-Century Ethnography, Literature, and Art. Cambridge, Mass: Harvard University Press.

Clifford, James, and *George E. Marcus, eds.* (1986): Writing Culture: The Poetics and Politics of Ethnography: A School of American Research Advanced Seminar. Berkeley: University of California Press.

Constitution (1936, 1936): Constitution (Fundamental Law) of the Union of Soviet Socialist Republics. http://www.departments.bucknell.edu/russian/const/36cons01.html, accessed March 11, 2013.

Constitution (1997, 1977): Constitution (Fundamental Law) of the Union of Soviet Socialist Republics., http://www.departments.bucknell.edu/russian/const/1977toc.html, accessed March 11, 2013.

Dean, Mitchell (1999): Risk, Calculable and Incalculable. In Risk and Sociocultural Theory. Deborah Lupton, ed. Pp. 131–159. Cambridge University Press.

Djelic, Marie-Laure, and *Sigrid Quack* (2010): Transnational Communities and Governance. In Transnational Communities: Shaping Global Economic Governance. Marie-Laure Djelic and Sigrid Quack, eds. Pp. 3–36. New York: Cambridge University Press.

Dobrovol'sky, Aleksandr, Yekaterina Petukhova, and *Igor' Karmazin* (2010): Перестройка, перестройка... ты за водкою постой-ка! Московский Комсомолец, July 5. http://www.mk.ru/social/article/2010/05/06/483636-perestroyka-perestroyka-tyi-za-vodkoyu-postoyka-foto.html, accessed April 8, 2013.

Dobrovol'sky, Vladimir (2009): Анатолий Быков выиграл дело против РФ в Страсбургском суде. РИА Новости. http://ria.ru/society/20090310/164333503.html, accessed July 31, 2013.

Douglas, Mary (1996): Purity and Danger: An Analysis of the Concepts of Pollution and Taboo. London: Routledge.

Douglas, Mary, and *Aaron B. Wildavsky* (1982): Risk and Culture: An Essay on the Selection of Technical and Environmental Dangers. Berkeley, Cal: University of California Press.

Dyatlov, Viktor I. (1996a): История «торговых меньшинств» на востоке. In Восток и Россия: взгляд из Сибири. Материалы и тезисы докладов к научно-практической конференции Pp. 316–320. Иркутск: Изд-во ИГУ.

–: (1996b): Предпринимательские меньшинства: торгаши, чужаки или посланные Богом? Симбиоз, конфликт, интеграция в странах Арабского Востока и Тропической Африки,. Москва: Наталис.

–: (2000): Современные Торговые Меньшинства: Фактор Стабильности Или Конфликта? (Китайцы И Кавказцы В Иркутске). Москва: Наталис.

–: (2009a) (ed.): Трансграничные миграции и принимающее общество: механизмы и практики взаимной адаптации. Екатеринбург: Изд-во Урал. ун-та.

–: (2009b): «Китайские рынки» российских городов – «уходящая натура». In Трансграничные миграции и принимающее общество: механизмы и практики взаимной адаптации. Viktor I. Dyatlov, ed. Pp. 249–288. Екатеринбург: Издательство Уральского университета.

Dyatlov, Viktor I., and *Konstantin V. Grigorichev, eds.* (2013): Переселенческое общество Азиатской России: миграции, пространства, сообщества. Иркутск: «Оттиск».

Dyatlov, Viktor I., and *Roman E. Kuznetsov* (2004): «Шанхай» в центре Иркутска. Экология китайского рынка. Экономическая социология 5(4 (Сентябрь)): 56–71.

Eder, M., A. Yakovlev, and *A. Çarkoglu* (2003): The Suitcase Trade Between Turkey and Russia: Microeconomics and Institutional Structure,. Working Paper WP4/2003/07 Moscow: State University — Higher School of Economics. Социология Рынков. http://www.hse.ru/data/467/989/1224/WP4_2003_07.pdf, accessed April 24, 2013.

Ewald, Francois (1991): Insurance and Risks. In The Foucault Effect: Studies in Governmentality. Graham Burchell, Colin Gordon, and Peter Miller, eds. Pp. 197–210. Chicago: The University of Chicago Press.

Foucault, Michel (1991a): Discipline and Punish: The Birth of the Prison. London: Penguin Books.

–: (1991b): Governmentality. In The Foucault Effect: Studies in Governmentality. Graham Burchell, Colin Gordon, and Peter Miller, eds. Pasquale Pasquino, tran. Pp. 87–107. Chicago: The University of Chicago Press.

Fromm, Erich (1968): Escape from Freedom. New York: Avon Books.

Füllsack, Manfred (2001): Official Figures and Unofficial Realities: Employment Rates and Their Significance in Russia. Europe-Asia Studies 53(4 June): 613–625.

Gallopin, Gary G. (2009): Beyond Perestroika: Axiology and the New Russian Entrepreneurs. Amsterdam: Rodopi.

Geertz, Clifford (1973): The Interpretation of Cultures: Selected Essays. New York: Basic Books.

–: (2000): Available Light: Anthropological Reflections on Philosophical Topics. Princeton, N.J: Princeton University Press.

Giddens, Anthony, and *Christopher Pierson* (1998): Conversations with Anthony Giddens: Making Sense of Modernity. Stanford, Calif.: Stanford University Press.

Gille, Zsuzsa (2010): Is There a Global Postsocialist Condition? Global Society 24(1): 9–30.

Golovanov, Dmitry (2012): В Красноярске родился миллионный житель. Российская газета. http://www.rg.ru/2012/04/10/reg-sibfo/millionnik.html, accessed October 7, 2014.

Gorbachev, Mikhail Sergeevich (1988): Perestroika: New Thinking for Our Country and the World. New, updated ed. London: Fontana.

Goskomstat (2013): Доля розничных торговых сетей в формировании оборота розничной торговли субъектов российской федерации. http://www.gks.ru/free_doc/new_site/business/torg/rozn/rozn26.htm, accessed December 2, 2014.

Granovetter, Mark S. (1973): The Strength of Weak Ties. American Journal of Sociology 78(6 May): 1360–1380.

–: (1985): Economic Action and Social Structure: The Problem of Embeddedness. American Journal of Sociology 91(3): 481–510.

Gudeman, Stephen (2010): Economic Anthropology. In The Routledge Encyclopedia of Social and Cultural Anthropology. Alan Barnard and Jonathan Spencer, eds. Pp. 210–216. London and New York: Routledge.

Gumuła, Wiesław (1990): Kilka uwag o transformacjach ustrojowych w Polsce w świetle "gospodarki cienia." In Gospodarka nieformalna: uwarunkowania lokalne i systemowe. Kazimierz Z. Sowa, ed. Pp. 235–250. Rzeszów: TNOiK.

Gurov, Aleksandr (1990): Профессиональная преступность: прошлое и современность. Москва: Юридическая литература. http://cl.rushkolnik.ru/docs/7840/index-49557.html, accessed July 3, 2013.

Gutmann, Peter M. (1977): The Subterranean Economy. Financial Analysts Journal: 26–28.

–: (1985): The Subterranean Economy, Redux. In The Economics of the Shadow Economy. W. Gaertner and A. Wenig, eds. Pp. 2–18. Berlin: Springer-Verlag.

Hann, Chris M. (2002a): Farewell to the Socialist "other." In Postsocialism: Ideals, Ideologies and Practices in Eurasia. Chris M. Hann, ed. Pp. 1–12. London: Routledge.

–: (2002b) (ed.): Postsocialism: Ideals, Ideologies and Practice in Eurasia. London: Routledge.

Hann, Chris M., and *Ildikó Hann* (1992): Samovars and Sex on Turkey's Russian Markets. Anthropology Today 8(4): 3–6.

Hann, Chris M., and *Keith Hart* (2011): Economic Anthropology: History, Ethnography, Critique. Cambridge: Polity.

Harding, Andrew (2011): How the Bull Fell Foul of the Russian Bear. The Guardian. www.theguardian.com/world/1999/nov/27/russia, accessed July 31, 2013.

Hart, Keith (1973): Informal Income Opportunities and Urban Employment in Ghana. The Journal of Modern African Studies 11(1): 61–89.

–: (2001): Money in an Unequal World. London: Textere.

–: (2011): Informal Economy. In The Human Economy: A Citizen's Guide. Keith Hart, Jean-Louis Laville, and Antonio David Cattani, eds. Pp. 142–153. Cambridge: Polity Press.

Hastrup, Kirsten (1995): A Passage to Anthropology: Between Experience and Theory. London: Routledge.

Hivon, Myriam (1998): The Bullied Farmer: Social Pressure as a Survival Strategy? In Surviving Post-Socialism: Local Strategies and Regional Responses in Eastern Europe and the Former Soviet Union. Susan Bridger and Frances Pine, eds. Pp. 33–51. London: Routledge.

Hobbes, Thomas (1991) [1651]: Leviathan. Cambridge Texts in the History of Political Thought. Cambridge: Cambridge University Press.

Humphrey, Caroline (1991): "Icebergs", Barter, and the Mafia in Provincial Russia. Anthropology Today: 8.

–: (2000): How Is Barter Done? The Social Relations of Barter in Provincial Russia. In The Vanishing Rouble: Barter Networks and Non-Monetary Transactions in Post-Soviet Societies. Paul Seabright, ed. Pp. 259–297. Cambridge: Cambridge University Press.

–: (2002a): Does the Category "postsocialist" Still Make Sense? In Postsocialism: Ideals, Ideologies and Practices in Eurasia Pp. 12–15. London: Routledge.

–: (2002b): The Unmaking of Soviet Life: Everyday Economies after Socialism. Culture & Society after Socialism. Ithaca: Cornell University Press.

Humphrey, Caroline, and *Vera Skvirskaja* (2009): Trading Places: Post-Socialist Container Markets and the City. Focaal – European Journal of Anthropology(55): 61–73.

IA Regnum (2013a): Губернатор Новосибирской области о закрытии Гусинобродского рынка: Это решение не подлежит пересмотру. http://www.regnum.ru/news/1741189.html, accessed August 20, 2014.

–: (2013b): В Новосибирске прошел масштабный митинг против закрытия Гусинобродской барахолки. http://www.regnum.ru/news/1743007.html, accessed August 20, 2014.

–: (2013c): Розничных рынков в Красноярском крае стало меньше. http://www.regnum.ru/news/1744281.html, accessed August 20, 2014.

Il'in, Vladimir I. (1994): Отечественный расизм. Рубеж: 189–204.

–: (2000): Классовая структура: проблема методологии анализа. Рубеж(15): 86–109.

–: (2008): Модели классообразования в пост-коммунистическом мире. Мир России 17(2). Journal Article: 3–21.

Ingold, Tim (2000): The Perception of the Environment: Essays on Livelihood,dwelling & Skill. London ; New York: Routledge.

Istomina, Natal'ya B. (2014): Славянский" рынок в Красноярске будут закрывать судебные приставы. http://afontovo.ru/news/14837, accessed February 28, 2014.

Ivleva, Irina V. (2001): Уличный рынок: среды петербургских торговцев. In Невидимые грани социальной реальности. Viktor Voronkov, Oleg Pachenkov, and Yelena Chikadze, eds. Pp. 83–95. Санкт-Петербург: ЦНСИ.

–: (2003): Петербургские торговцы: пример стратегии преодоления бедности в российском мегаполисе. Журнал социологии и социальной антропологии VI(Спецвыпуск): 284–304.

–: (2008): Трудовая деятельность и отношение к труду в «уличной экономике». Журнал социологии и социальной антропологии XI(3): 65–76.

–: (2009): Трудовые Мигранты В Городской Экономике. Журнал Социологии И Социальной Антропологии XII(3): 128–149.

–: (2010): Уличная экономика в повседневности переходного периода. Санкт-Петербург: Издательство Санкт-Петербургского ун-та.

Jarvie, Ian C., and *Joseph Agassi* (2010): Rationality. In The Routledge Encyclopedia of Social and Cultural Anthropology. Alan Barnard and Jonathan Spencer, eds. Pp. 590–594. New York: Routledge.

Jarzyńska, Katarzyna (2013): Ethnic Riots in Moscow. http://www.osw.waw.pl/en/publikacje/analyses/2013–10–16/ethnic-riots-moscow, accessed April 19, 2013.

–: (2014): Russia Tightens up Residence Regulations for CIS Citizens. OSW. http://www.osw.waw.pl/en/publikacje/analyses/2014–01–15/russia-tightens-residence-regulations-cis-citizens, accessed September 3, 2014.

Kapralova, Nina L., and *Larisa A. Karaseva* (2005): Челночный бизнес в российской экономике: роль и оценка масштабов. Экономический Журнал ВШЭ(3): 400–411.

Kapustin, Fyodor (2000): Последнее дело алюуминиевого короля. Труд, October 18. http://www.trud.ru/article/18–10–2000/13723_poslednee_delo_aljuminievogo_korolja.html, accessed July 31, 2013.

Katsenelinboigen, Aron (1977): Coloured Markets in the Soviet Union. Soviet Studies 29(1 (January)): 62–85.

Kieniewicz, Jan (2008): Ekspansja, kolonializm, cywilizacja. Warszawa: Wydawnictwo DiG and IBI AL.

Klyamkin, Igor, and *Lev Timofeyev* (2000): Теневая Россия. Москва: РГГУ.

Knight, Frank Hyneman (1964) [1921]: Risk, Uncertainty and Profit. New York: Reprints of Economic Classics.

Kokh, Alfred (2011): Некоторые мифы о реформах 90-х. http://lectures.gaidarfund.ru/articles/1148, accessed April 16, 2013.

Komarov, Pavel (2013): Гусинобродский рынок в Новосибирске: от "Черкизона" к Бирюлево. http://ria.ru/nsk/20131114/976868263.html, accessed August 20, 2014.

Komlev, Nikolay G. (2006): Словарь иностранных слов. http://dic.academic.ru/dic.nsf/dic_fwords/, accessed January 2, 2013.

Kornai, János (1980): Economics of Shortage. Contributions to Economic Analysis, 131. Amsterdam: North-Holland.

Kovács, János Mátyás (1992): Engineers of the Transition (Interventionist Temptations in Eastern European Economic Thought). Acta Oeconomica 44(1–2): 37–52.

–: (1997): Boxing the Shadow? "Neoliberals" and Social Quality After Communism. In The Social Quality of Europe. Beck Wolfgang, Laurent van Maesen, and Alan Walker, eds. Pp. 207–215. The Hague: Kluwer Law International.

Kovács, János Mátyás, and *Violetta Zentai* (2012): Prologue: Going beyond Homo Sovieticus. In Capitalism from Outside? Economic Cultures in Eastern Europe after 1989. János Mátyás Kovács and Violetta Zentai, eds. Pp. 1–14. Budapest: Central European University Press.

Koval'skaya, Mariya N. (2010): Бизнес по-китайски: трудовая деятельность китайских мигрантов на страницах иркутских газет. In Миграции и диаспоры в социокультурном, политическом и экономическом пространстве Сибири. Рубежи XIX–XX и XX–XXI веков. Viktor I. Dyatlov, ed. Pp. 263–281. Иркутск: Оттиск.

Kozłowski, Paweł (2004): Gospodarka nieformalna w Polsce. Dynamika i funkcje instytucji. Warszawa: PAN.

Krzyworzeka, Amanda (2013): Decision-Making in Farming Households in Eastern Poland. Focaal—Journal of Global and Historical Anthropology 2013(65): 129–144.

Kuznetsov, Sergey A. (1998): Толковый словарь русского языка. http://dic.academic.ru/contents.nsf/kuznetsov/, accessed December 28, 2012.

Lane, David Stuart (1992): Soviet Society under Perestroika. Completely rev. ed. London: Routledge.

Larin, Aleksandr (2008): Сколько в России китайских мигрантов? http://demoscope.ru/weekly/2008/0347/tema01.php#_FNR_1, accessed April 23, 2014.

Latour, Bruno (1993): We Have Never Been Modern. Cambridge, Mass: Harvard University Press.

Ledeneva, Alena V. (1998): Russia's Economy of Favours: Blat, Networking, and Informal Exchanges. Cambridge Russian, Soviet and Post-Soviet Studies, 102. New York: Cambridge University Press.

Lemon, Alaina (1998): "Your Eyes Are Green like Dollars": Counterfeit Cash, National Substance, and Currency Apartheid in 1990s Russia. Cultural Anthropology 13(1 February): 22–55.

Leybin, Vitaly (2006): Откуда пошли реформаторы. Интервью с Егором Гайдаром. http://www.polit.ru/article/2006/09/06/gaidar/, accessed April 18, 2013.

Lindley, Dennis V. (2006): Understanding Uncertainty. Hoboken, N.J.: Wiley-Interscience.

Lindquist, Galina (2000): In Search Of The Magical Flow: Magic And Market In Contemporary Russia. Urban Anthropology 29(4): 315–357.

–: (2006): Conjuring Hope: Magic and Healing in Contemporary Russia. Epistemologies of Healing, v. 1. New York: Berghahn Books.

Lin, Nan (2001): Building a Network Theory of Social Capita. In Social Capital: Theory and Research. Nan Lin, Karen Cook, and Ronald S. Burt, eds. Pp. 3–29. New York: Aldine de Gruyter.

Luhmann, Niklas (1993): Risk: A Sociological Theory. Communication and Social Order. Berlin: W. De Gruyter.

MacKenzie, David (1999): A History of Russia, the Soviet Union, and beyond. 5th ed. Belmont, CA: West/Wadsworth.

Malinowski, Bronislaw 2005 [1922]: Argonauts of the Western Pacific. An Account of Native Enterprise and Adventure in the Archipelagoes of Melanesian New Guinea. London: Routledge.

Mandel, Ruth, and *Caroline Humphrey, eds.* (2002): Markets and Moralities: Ethnographies of Postsocialism. Oxford: Berg.

Marciniak, Włodzimierz (2001): Rozgrabione imperium: upadek Związku Sowieckiego i powstanie Federacji Rosyjskiej. Kraków: Arcana.

Marx, Karl (1875): Critique of the Gotha Program. http://www.marxists.org/archive/marx/works/1875/gotha/ch01.htm, accessed March 11, 2013.

Mel'nichenko, Tatyana, Alberto Bolonini, and *Roberto Zavatta* (1997): Российский челночный бизнес. Общая характеристика и взаимосвязь с итальянским рынком. http://www.nisse.ru/business/article/article_1374.html?effort=, accessed May 15, 2013.

Minpromtorg Rossii (2013): Минпромторг России вводит санкции за нарушение требований к зданиям розничных рынков. http://old.minpromtorg.gov.ru/trade/hometrade/247/?print=1, accessed August 17, 2014.

Von Mises, Ludwig (1996) [1949]: Human Action: A Treatise on Economics. San Francisco: Fox & Wilkes.

Mokrzycki, Edmund (1996): A New Middle Class? In Culture, Modernity and Revolution: Essays in Honour of Zygmunt Buman. Richard Kilminster and Ian Varcoe, eds. Pp. 184–200. London: Routledge.

Morrison, John (1991): Boris Yeltsin: From Bolshevik to Democrat. New York: Dutton.

Mörtenböck, Peter, and *Helge Mooshammer* 2007: Trading Indeterminacy: Informal Markets in Europe. Field: A Free Journal for Architecture, vol. 1 (September): 73–87.

Moshkov, Aleksandr (2013): Рынки раздавят буквой закона. http://www.vegaslex.ru/text/65815, accessed August 17, 2014.

Nazpary, Joma (2002): Post-Soviet Chaos: Violence and Dispossession in Kazakhstan. London: Pluto Press.

Nechayev, Andrey (2011): Предотвращенная катастрофа,. http://lectures.gaidarfund.ru/articles/1146, accessed April 16, 2013.

Nowak, Stefan (1979): System wartości społeczeństwa polskiego. Studia socjologiczne(4): 155–173.

O'Malley, Pat (2008): Governmentality and Risk. In Social Theories of Risk and Uncertainty: An Introduction. Jens Zinn, ed. Pp. 55–75. Malden, MA: Blackwell Pub.

Orlov, Aleksey (2013): Враньём сыт не будешь. Газета Городские новости+, March 9.

Overchenko, Mikhail, and *Dmitry Dmitriyenko* (2011): Абрамович признал, что аукцион по «Сибнефти» был фикцией. http://www.vedomosti.ru/politics/news/1410605/abramovich_priznal_chto_aukcion_po_sibnetfi_byl_fikciej, accessed July 25, 2013.

Owen, Thomas C. (1995): Russian Corporate Capitalism from Peter the Great to Perestroika. New York: Oxford University Press.

Ozhegov, Sergey I., and *Natalya Yu. Shvedova* (1949): Толковый словарь. http://dic.academic.ru/dic.nsf/ogegova, accessed December 28, 2012.

Pavlovsky, Gleb (2014): Putin's World Outlook 88: 54–66.

Pawelczyk, Grzegorz, and *Piotr Kudzia* (2001): Stadion w państwie. Wprost, June 17. http://www.wprost.pl/ar/?O=10285, accessed April 19, 2011.

PenzaNews (2011): От переезда рынков под крышу не пострадают ни население, ни предприниматели – Нина Богач. http://penzanews.ru/interviews/53635-2011, accessed August 17, 2014.

Pikhoya, Rudolf G., and *Lev Timofeyev* (1999): Судьбы реформ и реформаторов в России. Москва: Издательство РАГС.

Polanyi, Karl (1957): The Economy As Instituted Process. In Trade and Market in the Early Empires: Economies in History and Theory. Karl Polanyi, Conrad M. Arensberg, and Harry W. Person, eds. Pp. 243–270. Glencoe, Illinois: The Free Press.

–: (2001): The Great Transformation. Beacon Paperbacks, 45. Boston: Beacon Press.

Polyakov, Vitaly (2014): В Красноярске начался снос "Славянского" рынка. http://afontovo.ru/news/15002, accessed March 6, 2014.

Popov, Vladimir (2013): "Because the Chinese Growth Model Became so Successful in Ensuring Catch-up Development It Has Become Extremely Appealing in the Developing World", Conversation with Piotr Dutkiewicz. In 22 Ideas to Fix the World. Conversations with the World's Foremost Thinkers. Piotr Dutkiewicz and Richard Sakwa, eds. Pp. 394–413. New York and London: New York University Press.

Poretskina, Yevgeniya (2006): "Челночный" бизнес. Краткая история вопроса и его особенности в Санкт-Петербурге. Телескоп: журнал социологических и маркетинговых исследований(5): 24–31.

Prikhod'ko, Sergey V., and *Nadezhda P. Volovik* (1997): Индивидуальный импорт: аспекты развития. Экономико-политическая ситуация в России(Сентябрь): 17–18.

Protsenko, Pavel (2013): Уничтожение российского рынка. Во имя чего? Ежедневный Журнал. http://www.ej.ru/?a=note&id=23899, accessed August 18, 2014.

Putnam, Robert D., Robert Leonardi, and *Raffaella Nanetti* (1993): Making Democracy Work: Civic Traditions in Modern Italy. Princeton, N.J: Princeton University Press.

Radayev (Radaev), Vadim (1998): Формирование новых российских рынков: транзакционные издержки, формы контроля и деловая этика. Москва: Центр политических технологий. http://ecsoc.ru/db/msg/9956/textlink, accessed July 29, 2013.

–: (1999): Неформальная экономика и внеконтрактные отношения в российском бизнесе. Подходы к исследованию неформальной экономики. In Неформальная экономика. Россия и мир. Teodor Shanin, ed. Pp. 35–60. Москва: Логос.

–: (2003): Изменение конкурентной ситуации на российских рынках (на примере розничных сетей). Социология рынков, WP4/2003/06. Москва: ГУ

ВШЭ. https://www.hse.ru/data/2010/05/05/1216427428/WP4_2003_06.pdf, accessed October 9, 2013.

–: (2005): Захват российских территорий: деловые стратегии розничных компаний в 2000'е гг. Социология рынков, WP4/2005/03. Москва: ГУ ВШЭ. https://www.hse.ru/data/2010/05/05/1216427469/WP4_2005_03.pdf, accessed October 9, 2013.

–: (2006): Что такое рынок: экономико-социологический подход. Социология рынков, WP4/2006/07. Москва: ГУ ВШЭ. https://www.hse.ru/data/2010/05/05/1216427487/WP4_2006_07.pdf, accessed October 9, 2013.

–: (2011a): Кому принадлежит власть на потребительских рынках: отношения розничных сетей и поставщиков в современной России. Москва: Издательский дом ВШЭ.

–: (2011b): Where Does the Demand for Regulation Come From? The State's Return to the Retail Trade in Russia. Higher School of Economics Research Paper (No. WP BRP 02/SOC/2011). Higher School of Economics Research Paper, Sociology. http://www.hse.ru/en/org/hse/wp/prepfr_SOC, accessed October 22, 2014.

Radzikhovsky, Leonid (1995): Номенклатура обменяла «Капитал» на капитал. Известия, July 3.

Ries, Nancy (1997): Russian Talk: Culture and Conversation during Perestroika. Ithaca, N.Y: Cornell University Press.

–: (2002): "Honest Bandits" and "Warped People": Russian Narratives about Money, Corruption, and Moral Decay. In Ethnography in Unstable Places: Everyday Lives in Contexts of Dramatic Political Change. Carol J. Greenhouse, Elisabeth Mertz, and Kay B. Warren, eds. Pp. 267–315. Durham: Duke University Press.

–: (2009): Potato Ontology: Surviving Postsocialism in Russia. Cultural Anthropology 24(2): 181–212.

Rogers, Douglas (2005): Moonshine, Money, and The Politics of Liquidity In Rural Russia. American Ethnologist 32(1): 63–81.

Rosefielde, Steven (2001): Premature Deaths: Russia's Radical Economic Transition in Soviet Perspective. Europe-Asia Studies 53(8 (December)): 1159–1176.

Rupnik, Jaques (2002): The Postcommunist Divide. In Democracy after Communism. Larry Diamond and Marc F. Plattner, eds. Pp. 103–108. Baltimore: Johns Hopkins University Press.

Sakwa, Richard (1999): The Rise and Fall of the Soviet Union, 1917–1991. Routledge Sources in History. London: Routledge.

Salecl, Renata (2010): Choice. London: Profile Books.

Samokhvalova, Yekaterina, and *Екатерина Самохвалова* (2014): Чиновники мэрии встретились с владельцами павильонов и автостоянок. http://afontovo.ru/news/16209, accessed October 22, 2014.

Sampson, Steven L. (1987): The Second Economy of the Soviet Union and Eastern Europe. Annals of the American Academy of Political and Social Science 493(September): 120–136.

Sarnov, Benedikt (2005): Наш советский новояз. Маленькая энциклопедия реального социализма. Москва: Эксто.

Schwirtz, Micheal (2008): Vory v Zakone Has Hallowed Place in Russian Criminal Lore, The New York Times. http://www.nytimes.com/2008/07/29/world/europe/29iht-moscow.4.14865004.html?pagewanted=all&_r=0, accessed July 3, 2013.

Seabright, Paul, ed. (2000): The Vanishing Rouble: Barter Networks and Non-Monetary Transactions in Post-Soviet Societies. Cambridge: Cambridge University Press.

Semikina, Anastsiya (2013): «Славянский рынок» снесут сразу после новогодних праздников. In. http://www.avtoritetnoeradio.ru/news/society/view/slavjanskij_rynok_snesut_srazu_posle_novogodnikh_prazdnikov, accessed March 1, 2014.

Shavlokhova, Madina (2014): «Если люди попросят нас о помощи». Владимир Путин впервые оценил последние события на Украине. http://www.vz.ru/politics/2014/3/4/675487.html, accessed December 2, 2014.

Sibnovosti (2014 a): В Красноярске уберут павильоны со всех знаковых мест. http://krsk.sibnovosti.ru/society/261032-v-krasnoyarske-uberut-pavilony-so-vseh-znakovyh-mest, accessed October 22, 2014.

–: (2014b): В Красноярске исчезнут незаконные рынки. http://krsk.sibnovosti.ru/society/261788-v-krasnoyarske-ischeznut-nezakonnye-rynki, accessed October 22, 2014.

Sik, Endre, and *Claire Wallace* (1999): The Development of Open-Air Markets in East-Central Europe. International Journal of Urban and Regional Research 23(4): 697–714.

Simmel, Georg (1971): The Stranger. In On Individuality and Social Forms. Selected Writings. Donald N. Levine, ed. Pp. 143–149. Chicago and London: The University of Chicago Press.

Smith, Adam (2007) [1776]: The Wealth of Nations. Petersfield: Harriman House.

Smorshchkov, Pavel (2013): Дефолт: 15 лет свободы рубля. http://www.gazeta.ru/business/2013/08/16/5574905.shtml, accessed December 17, 2013.

Snisarenko, A. (1999): Этническое предпринимательство в большом городе современной России (на примере исследования азербайджанской общины в Петербурге. In Неформальная экономика. Россия и мир. Teodor Shanin, ed. Pp. 138–155. Москва: Логос. http://www.narcom.ru/publ/info/522, accessed April 16, 2014.

Sowa, Kazimierz Z. (1990): Gospodarka nieformalna: uwarunkowania lokalne i systemowe. Rzeszów: TNOiK.

Stoner-Weiss, Kathryn (2006): Resisting the State: Reform and Retrenchment in Post-Soviet Russia. Cambridge: Cambridge University Press.

Sztompka, Piotr (1999): Trust: A Sociological Theory. Cambridge Cultural Social Studies. Cambridge: Cambridge University Press.

–: (2000): Trauma Wielkiej Zmiany. Społeczne Koszty Transformacji. Warszawa: ISP PAN.

Tarascio, Vincent J. (1985): Cantillon's "Essai": A Current Perspective. Journal of Libertarian Studies VII(7 Fall): 249–257.

Telegina, N. W. (2012): Ответы на вопросы по деятельности рынков в г. Красноярске. http://citizengrad.narod.ru/nemolchim.files/otveti.htm, accessed August 20, 2014.

Timofeyev, Lev (2000): Институциональная коррупция. Москва: РГГУ.

–: (2008): Теневые экономические системы современной России. Москва: РГГУ.

Tischner, Józef (1992): Etyka solidarności oraz Homo sovieticus. Kraków: Znak.

Tyrmand, Leopold (2009): Dziennik 1954. Warszawa: Wydawnictwo MG.

Varese, Federico (2001): The Russian Mafia: Private Protection in a New Market Economy. Oxford, England: Oxford University Press.

Varnavsky, Pavel K. (2013): Кто этнизирует мигрантов: принимающее общество или они сами? Институционализация мигрантского сообщества на примере выходцев из Кыргызстана в Улан-Удэ. In Переселенческое общество Азиатской России: миграции, пространства, сообщества. Viktor I. Dyatlov and Konstantin V. Grigorichev, eds. Pp. 466–491. Иркутск: », pp 466–491.: Оттиск.

Vasin, Sergey A. (2008): Возрастной состав населения России. In Население России 2006: Четырнадцатый ежегодный демографический доклад. Anatoly G. Vishnevsky, ed. Pp. 47–70. Москва: Издательский дом ГУ-ВШЭ.

Verdery, Katherine (1996): What Was Socialism, and What Comes Next? Princeton Studies in Culture/power/history. Princeton, NJ: Princeton University Press.

–: (1997): Fuzzy Property: Rights, Power, and Identity in Transylvania's Decollectivization. In Transforming Post-Communist Political Economies. Joan M. Nelson, Charles Tilly, and Lee Walker, eds. Pp. 102–117. Washington, D.C: National Academy Press.

Vestifinance (2014): ФНС: количество ИП в 2013 году резко сократилось. vestifinance.ru. http://www.vestifinance.ru/articles/38459, accessed August 20, 2014.

Vorkozhova, Tatyana (2013): "Славянский базар". Громкое дело. http://www.iapress-line.ru/lenta-dnya/item/31392-slavyanski-rynok, accessed March 1, 2014.

Vyazemsky, Pyotr (1963): Записные Книжки. Москва: Издательство Академии Наук СССР.

Weber, Max (1991) [1919]: Science as a Vocation. In From Max Weber: Essays in Sociology. H. H. Gerth and C. Wright Mills, eds. H. H. Gerth and C. Wright Mills, trans. Pp. 129–156. London: Routledge & Kegan Paul.

–: (2001) [1905]: The Protestant Ethic and the Spirit of Capitalism. Talcott Parsons, tran. Routledge Classics. London and New York: Routledge.

Wedel, Janine R. (1986): The Private Poland. New York, NY: Facts on File.

Wegren, Stephen K., David J. O'Brien, and *Valeri V. Patsiorkovski* (2003): Russia's Rural Unemployed. Europe-Asia Studies 55(6 (September_): 847–867.

Yakovlev, Andrey A., Viktoriya V. Golikova, Nina L. Kapralova, et al. (2007): Российские «челноки»–от предпринимателей поневоле к интеграции в рыночное хозяйство. Мир России 16(2): 84–106.

Yurchak, Alexei (2002): Entrepreneurial Governmentality in Postsocialist Russia: A Cultural Investigation of Business Practices. *In* The New Entrepreneurs of Europe and Asia: Patterns of Business Development in Russia, Eastern Europe and China. Victoria E. Bonnell and Thomas B. Gold, eds. Pp. 278–324. New York: M.E. Sharpe.

–: (2006): Everything Was Forever, until It Was No More: The Last Soviet Generation. In-Formation Series. Princeton, NJ: Princeton University Press.

Zakon o razvitii malogo i srednego predprinimatel'stva (2007): Федеральный закон от 24.07.2007 N 209-ФЗ "О развитии малого и среднего предпринимательства в Российской Федерации" (с изм. и доп., вступ. в силу с 01.07.2014). http://www.consultant.ru/document/cons_doc_LAW_157188/, accessed August 22, 2014.

Zakon o roznichnykh rynkakh (2006a): Федеральный закон от 30 декабря 2006: г. N 271-ФЗ "О розничных рынках и о внесении изменений в Трудовой кодекс Российской Федерации." http://base.garant.ru/190400/1/#block_1, accessed February 25, 2014.

–: (2006b): Федеральный Закон От 30 Декабря 2006 Г. N 271-ФЗ "О Розничных Рынках И О Внесении Изменений В Трудовой Кодекс Российской Федерации." the law available at: http://base.garant.ru/190400/1/#block_1, accessed February 25, 2014.

Zarakhovich, Yuri (2006): Inside Russia's Racism Problem. Time. http://content.time.com/time/nation/article/0,8599,1304096,00.html, accessed March 3, 2014.

Zavisca, Jane (2003): Contesting Capitalism at the Post-Soviet Dacha: The Meaning of Food Cultivation for Urban Russians. Slavic Review 62(4 (Winter)): 786–810.

Zhou, Min (2004): Revisiting Ethnic Entrepreneurship: Convergences, Controversies, and Conceptual Advancements. International Migration Review 38(3 (Fall)): 1040–1074.

Zinn, Jens (2008): Introduction: The Contribution of Sociology to the Discourse on Risk and Uncertainty. In Social Theories of Risk and Uncertainty: An Introduction. Jens Zinn, ed. Pp. 1–17. Malden, MA: Blackwell Pub.

Zinovyev (Zinoviev), Aleksandr A. (1980): The Radiant Future. 1st Trade ed. New York: Random House.

–: (1981): Мы и Запад. Стати, интервью, выступления 1979–1980 гг. Lausanne: L'Age d'homme.

–: (1982): Homo Sovieticus. Charles Janson, tran. Boston: Atlantic Monthly Press.

–: (1988): Горбачебизм. New York: Liberty Publishing House.

–: (1990): Катастройка: повест' о перестройке в Партграде. Lausanne: L'Age d'homme.

–: (1994): Коммунизм как реальность; Кризис коммунизма. Москва: Центрполиграф.

–: (2005): Русская трагедия. Москва: Издательство Эксто.

Zyblyev, Sergey (2012): В Красноярске родился миллионный житель. http://www.vesti.ru/doc.html?id=768106&cid=7, accessed October 7, 2014.

Photographies

1. The entrance to the Slavyansky market. The notice says: *The Zheleznodorozhny District Fair.* Calling the Slavyansky a *fair* not a *market* was a juridical trick to prolong its existence.

2. Shopping at the Slavyansky.

3. The clothes branch at the Slavyansky was the domain of Kyrgyzs.

4. Some of the tradespeople still work outside the entire year.

5. An Uzbek trader at his spices stand.

6. This is the way Slavyansky traders got informed about what would happen to their market. The billboard displays the project of the new shopping mall.

7. The remains of the GorDK market. One can see a new housing estate in construction.

8. The Vostochny market is a part of the huge Krastets. Substantial parts of the Vostochny are owned by the Chinese.

9. The unroofed part of the Krastets. The chimneys in the background belong to the Krasnoyarsk Heat and Power Plant (KrasTETs), which gave the name to the market.

10. The Krastets on a windy and rainy day

11. Throughout the Krastets new shopping malls have been constructed

12. The neighbourhood of the Krastets. The billboard offers juridical support in procuring permits for temporary stay and acquiring Russian citizenship. Bus stops in Krasnoyarsk are literally plastered with similar ads

13. The entrance to the Tsentral'ny market. The chapel on the hill at the right was built in a supposed geographical centre of Russia

14. The Tsentral'ny was a busy market once. All the shops at this used to be open. In the middle, one can also still see the lines that separated now non-existent stalls

15. Mira street in the city centre becomes a fair for a few days a year

16. A stand at the fair

17. An Azerbaijani trader in front of his kiosk

18. A street stand with sausages

19. There were so-called social places at the Slavyansky, where pensioners could sell their produce free of charge

20. Some pensioners, however, preferred to sell at more busy places. They sold informally next to a supermarket

Index

A
Andropov, Yury 41
Afanasyev, Yury 54, 65, 80, 117, 126, 153
Alekperov, Vagit 53
Aslund, Anders 47–54, 57, 59, 67, 73, 80, 82, 83, 153
Aven, Pyotr 67, 157
Aristotle 32, 33, 153

B
Beck, Ulrich 15, 26, 154
Beck-Gernsheim, Ursula 15, 26, 154
Berezovsky, Boris 53, 80, 150
Brednikova, Ol'ga 90, 103, 104, 106, 110, 111, 154
Brezhnev, Leonid 41, 42, 47, 49, 53
Buchowski, Michał 9, 14, 15, 22, 41, 77, 126, 155
Burawoy, Michael 13, 27, 30, 37, 58, 60, 61, 64, 117, 118, 125, 132–134, 155
Bykov, Anatoly 86, 87, 95

C
Cherkasov, Anatoly 87
Chernyshevsky, Nikolay 18, 19, 59, 155
Chłopecki, Jerzy 38, 40, 155
Chubais, Anatoly 80

D
Douglas, Mary 16, 26, 156, 157
Dyatlov, Viktor 9, 89, 90, 102, 104–107, 109, 154, 157, 161, 166

F
Foucault, Michel 17, 18, 155, 157

G
Gaidar, Yegor 11, 23, 36, 55, 57, 58, 67, 71
Geertz, Clifford 125, 126, 133, 157
Gorbachev, Mikhail 7, 11, 36, 47–49, 53, 54, 57, 60, 153, 158
Grigorichev, Konstantin 90, 102, 104, 106, 157, 166
Gusinsky, Vladimir 80

H
Hann, Chris 14, 21, 27, 31–33, 64, 70, 72, 126, 158
Hart, Keith 14, 27, 31–33, 36, 41, 64, 158, 159
Humphrey, Caroline 14, 16, 21, 37, 53, 55, 56, 59, 60, 64, 79, 85, 117, 121, 159, 162

K
Khodorkovsky, Mikhail 53, 80
Khrushchev, Nikita 41
Kieniewicz, Jan 9, 22, 160
Klyamkin, Igor 36, 37, 80, 86, 121, 160
Knight, Frank 25, 28, 29, 31, 160
Korchagin, Pavel 43
Kornai, János 37, 40, 160
Kuznetsov, Roman 39, 44, 90, 157, 161

L
Latour, Bruno 133, 161
Ledeneva, Alena 53, 54, 161
Lenin, Vladimir 47
Lozishvili, Otari 46

M
Malinowski, Bronisław 30, 162
Marx, Karl 42, 63, 162
Medvedev, Dmitry 16, 91
Mill, John Stuart 31

N
Nechayev, Andrey 55, 57, 67, 75, 162

P
Pachenkov, Oleg 90, 103, 154, 160
Pascal, Blaise 28
Pavlov, Valentin 57, 58
Polanyi, Karl 26, 32, 89, 163
Potanin, Vladimir 53
Prokhorov, Mikhail 53
Putin, Vladimir 66, 95, 117, 144, 163

R
Radayev (Radaev), Vadim 11, 20, 38, 74, 82, 118, 163
Radzikhovsky, Leonid 54, 164
Rogers, Douglas 64, 65, 164

S
Saltykov-Shchedrin, Mikhail 91
Savchuk, Yury 96
Shatalin, Stanislav 57
Smith, Adam 32, 165
Stakhanov, Alexei 43
Stalin, Joseph 35, 42, 44, 49, 54
Sztompka, Piotr 35, 41, 43, 44, 90, 122, 165

T
Timofeyev, Lev 36, 37, 54, 80, 86, 121, 160, 163, 165
Tischner, Józef 35, 41, 43, 166

V
Varese, Federico 79, 80, 82, 166
Varnavsky, Pavel 90, 103, 104, 166
Vyazemsky, Pyotr 91, 92, 166

W
Weber, Max 27, 40, 166
Wedel, Janine 38, 39, 43, 46, 166

Y
Yavlinsky, Grigory 57
Yeltishev, Vasily 93, 95
Yeltsin, Boris 48, 58, 60, 67, 150, 162
Yurchak, Alexei 35, 36, 39, 55, 166

Z
Zinovyev (Zinoviev), Aleksandr 11, 27, 41–43, 48, 54, 84, 167

www.ingramcontent.com/pod-product-compliance
Ingram Content Group UK Ltd.
Pitfield, Milton Keynes, MK11 3LW, UK
UKHW041831100926
8556IPUK00002B/120

9 783631 668559